"Many of us believe that while Generative AI can improve our efficiency, it comes at the cost of losing our authenticity. In this candid and wonderful book, Allison Shapira offers a compelling vision and practical advice for how we can use AI in ways that not only prevent us from losing authenticity but help us bring out our authentic selves. I highly recommend this book to any leader seeking to communicate more effectively, efficiently, and authentically with the help of AI."

—Dan Levy, Faculty, Harvard Kennedy School and Co-Author, *Teaching Effectively with ChatGPT*

"As AI capabilities and impact accelerate, we need leaders like Allison who remind us of the power of human connection and authenticity. *AI for the Authentic Leader* is a timely and essential book on one of the most fundamental questions to the future of work, education and business: How can AI make us more human?"

—Paul Roetzer, Founder and CEO, SmarterX

"Allison Shapira has written an AI book that puts humans first. I've seen firsthand how easy it is for business leaders to lose sight of authenticity in the rush toward automation. Allison cuts through the complexity with practical guidance that helps leaders use AI to amplify their authenticity, not replace it. Her AI Authenticity Loop is exactly the timely roadmap executives need for harnessing AI's power while staying true to themselves and their people."

—Matt Strain, Founder, The Prompt

"*AI for the Authentic Leader* hits the sweet spot between tech innovation and real-world leadership. Allison Shapira strips away the hype and delivers clear, actionable strategies that help leaders use

AI to communicate with impact—without losing their authenticity. If you're serious about staying relevant, increasing your influence, and leading effectively in the age of AI, this book belongs in your toolkit."

—Ford Saeks, CEO Prime Concepts Group, Inc.

"Allison Shapira pulls no punches and lets you know exactly what is going on with AI and how to be authentic and stand out. This book doesn't just inspire you (it certainly does that), but it also ignites the desire to be authentic in an AI-dominant world. This guide is indispensable if you want to succeed today and tomorrow. If you're a communicator who writes, uses video, uses audio only, or is trying to make sense of the wacky, crazy world of AI, this is a must-read. Get this book and read it, yes. But go beyond that to use and implement its recommendations. Doing that, you'll stand out in a profoundly authentic way. You'll love it!"

—Terry Brock, Co-Founder, Stark Raving Entrepreneurs

"As someone who has scaled million-dollar wealth management teams for over twenty years, I've seen how authentic communication drives results. This book brilliantly shows leaders how to leverage AI while maintaining the genuine connection that builds lasting client relationships."

—Antoinette Rodriguez, MBA, 20-plus year
Wall Street veteran and 4X Chairwoman of
Financial Advisor magazine's Invest in Women

From *The Washington Post* bestselling
author of *Speak with Impact*

Ai

FOR THE

AUTHENTIC

LEADER

How to Communicate More Effectively
without Losing Your Humanity

ALLISON SHAPIRA

SPA CREEK
PRESS

Published by Spa Creek Press

SPA CREEK
PRESS

Produced by GMK Writing and Editing, Inc.

Managing Editor: Katie Benoit

Copyedited by: Amy Gordon

Proofread by: Elizabeth Crooks

Cover Design by: Vicky Vaughn Shea

Interior Design by: Vicky Vaughn Shea

Interior Layout by: Joanna Beyer

Print ISBN: 978-1-966981-15-2

Ebook EISN: 978-1-966981-16-9

Visit the author's website at https://www.allisonshapira.com

Write the author at info@globalpublicspeaking.com

This work was written, edited, designed, and produced by humans. While AI tools were occasionally consulted during the process, the content and production was completed through the efforts of human authorship and publishing professionals.

Acknowledgments

Generative AI is an emerging field that evolves weekly. The knowledge level of readers varies widely, as does their readiness to engage with it. I am grateful to a number of trusted friends and colleagues who provided feedback to ensure this book would resonate across all audiences. Matt Strain, my good friend and founder of The Prompt, was the ultimate thought partner—always willing to dive into the newest AI developments and discuss their implications for business leaders. Dr. Allison Sadler, fellow author and head of global learning at a multinational company, provided nuanced feedback based on her deep leadership expertise. Sabra Horne, with her detailed editorial eye and extensive publishing experience, helped fine-tune the book during the final stretch of the writing process. Nick Unger, founder of AlterityAI, gave strategic feedback on the text of the keynote speech that ultimately became the first draft of this book. Fellow speakers and authors Bill Cates and Ford Saeks were invaluable: Bill was an important accountability partner and Ford's book on AI inspired me to write my own. Alan Weiss, business advisor, provided the strategic guidance to get it done. Veronica Pirillo, whose authentic cover letter cut through the AI-written crowd, provided incredible support throughout the writing process. Finally, thank you to Sarah Turkel, Cara Paulan, and Kim Fredrich, whose creativity and tough love during our weekly meetings created the conditions where the AI Authenticity Loop was formed.

Contents

The Human Connection Imperative

A Hush Fell Over the Crowd

There's an energy that humans create when we gather together in person. It comprises each person's attention, expectations, and anticipation for what comes next. This is what happens when we say, "A hush fell over the crowd" or when everyone in a sports stadium does The Wave.

There is a name for this phenomenon. It's called *collective effervescence*. Coined by French sociologist Émile Durkheim in his book *The Elementary Forms of the Religious Life*, it is the shared harmony created by group rituals. We feel it during religious services, musical concerts, community rallies, sporting events, and even professional conferences. It creates a sense of community and connection.

We humans became the dominant species on this planet not because we were the fastest or largest animals, but because of our ability to collaborate across large communities. Communication played a critical role in this process. The historian and philosopher Yuval Noah Harari has said that "storytelling . . . the ability

to imagine narratives, is the driving force of history and the ...
unique superpower of homo sapiens." Community and connection
have allowed us to thrive as a species.

How I Learned to Create Connection

For the past thirty years, I have experienced this feeling of connection—this collective effervescence—as a performer, professional
speaker, and executive advisor.

Training to be an opera singer, I observed the impact that clas-
sical music had on my audience. There is something otherworldly
about an opera singer's command of her voice that, to the lay
person, feels like magic and can move them to tears. Later in life,
when I learned to play the guitar and started singing folk music, I
saw a different effect. This musical genre made my audience lean
in and sing along. In fact, throughout history, folk music has played
an important role in bringing people together both socially and
politically.

In his book *Stage Performance*, the musician Livingston Taylor
wrote, "The job of a stage performer is to invite the audience to
leave their reality and enter yours." I would say that the job of a
performer—and speaker—is to *co-create a new reality between
you and everyone in the room*. This is what charismatic speakers
and leaders do on stage. In fact, studies of interaction ritual theory
demonstrate the powerful effect it has on creating shared bonds.

My career has taken me from opera to diplomacy and from
academia to business. Every step of the journey has focused on
bringing humans together. For over twenty years, I have advised
thousands of leaders around the world, through my keynote
speeches, executive workshops, or one-on-one executive advisory.

Together, we work on building their confidence in order to be more effective leaders.

Why Every Leader Needs This Now

The leaders I advise—many of them corporate executives—have realized the more senior they become, the more time they spend communicating the messages of their organization. How you communicate is how you lead.

Through this work, I developed the ACE Model of Leadership Communication, recognizing that everything I was teaching came back to the same three core components: authenticity, clarity, and energy. No matter what country you live in, what language you speak, or in what industry you work, every one of us needs these three components in order to speak with impact (the subject of my first book) and be a more effective leader. This book will focus on the relationship between AI and authenticity.

Why AI Caught My Attention

For the past eight years, tech entrepreneurs have approached me to test their public speaking apps and websites. More and more, these platforms started incorporating artificial intelligence. In using these tools, I realized AI's potential to improve our communication skills: It can help us identify and reduce our filler words like *um* and *ah*, provide feedback on our pacing and pausing, and help us achieve clarity.

But my work goes beyond the technical mechanics of presentation skills. What about communication at its most profound level: Could AI improve our authenticity and therefore our ability to connect with one another? Could AI make us better leaders?

These questions piqued my curiosity. It is precisely because of my focus on leadership and human connection that I became fascinated by the nonhuman tools that have entered our working and living spaces.

Throughout human evolution, we have invented new ways to connect with one another, such as the steam engine, the printing press, and the Internet. Each invention facilitated human connection in ways that were previously impossible, whether physical, intellectual, or emotional.

The AI Evolution You Can't Ignore

We are now living through an even greater evolution, which brings with it an untold number of new ways we can connect—if we harness it effectively. How will AI affect our human interactions? In which ways can it help us be better humans? Can it amplify our ability to connect with one another? These are the questions I will discuss in this book.

I believe AI will transform every part of our lives—for better *and* for worse. We have had AI in our lives for decades. It started in the 1950s and has gone in and out of fashion. It is all around you whether or not you realize it: part of your smart vacuum, your digital devices, and your online shopping recommendations.

The field of AI and machine learning is a vast and evolving landscape. Traditionally, AI systems were rule based: You programmed them to follow specific instructions to accomplish a set of tasks, such as play chess or sort data. Then, machine learning enabled models to learn from patterns in data. This book will focus on *generative* AI—tools that produce new content: text, images, music, and insights. While generative AI is only a small subset of the field of AI, it is the one you have the most direct access to, ever since the organization OpenAI made ChatGPT widely available in November 2022.

What Is Artificial Intelligence?

I define artificial intelligence as the ability of machines to carry out tasks we previously thought only humans could perform. Today, AI is increasingly capable of tasks we thought were unique to humans. In my keynote speeches on AI and authenticity, I often ask the audience, "What does it mean to be human?" The responses invariably include such characteristics as empathy, creativity, and strategic thinking. But now, AI excels at demonstrating those qualities. When we can train AI chatbots to be empathetic or create algorithms that creatively connect the dots among disparate topics, we narrow the list of competencies that make us human. You could say that AI is challenging what it means to be human. We are all at the start of a collective exploration, and in this book I will share my learnings as I journey down that path.

I am not an expert in AI. I write from the perspective of someone who has devoted her entire career to helping humans connect with one another in a physical or virtual environment—from music to communication to leadership. My interdisciplinary perspective spans different sectors—having worked in government, nonprofit, and the private sector—as well as different countries, having studied ten foreign languages.

We cannot leave the future to AI experts: This is a general-purpose technology that will affect everyone, and so we need philosophers, sociologists, historians, artists, scientists, doctors, and experts from all fields to come together to share their knowledge. This book is my own first contribution to the field.

Your AI Strategy Starts Here

This book is a practical guide, a cautionary warning, and a series of reflections. The first section focuses on how to strategically use artificial intelligence to help you tap into your authenticity in a way that makes you a more effective leader, especially in the way you communicate. It does not teach you how to write the perfect prompt or use the latest AI tools—there are already many resources on those topics, and they will all change as the tools evolve.

This book provides ideas about how to use AI to be a *more* authentic leader, not less. As opposed to outsourcing your writing or your strategy to AI, learn to use AI to bring out your own ideas, thoughts, and voice. Learn how to use AI to become a better human, in a way that sharpens your instincts and helps you lead with integrity and purpose. You will learn strategies to help you overcome imposter syndrome, quiet the self-doubt in your mind, and speak with executive presence. You will learn my framework, the AI Authenticity Loop, to guide you when using AI to communicate an important message.

The second section focuses on cautions, concerns, insights, and questions about how we need to prepare for the future. You will learn about my biggest fears related to AI and how we need to protect ourselves from the misuse of AI by bad actors. The decisions we make about AI today and over the next few years will affect the trajectory of human development. We need to play an active role in those decisions. The book will conclude with a look at how, together, we can prepare for the future of human connection.

AI as Your Co-Pilot: Augmenting Human Potential

My goal is to help you augment your human capabilities, not replace them. I think of AI tools not as a self-driving car, but as the driver-assist function in your car. You remain in control, while having an ongoing presence that helps you focus. Alternatively, think of AI as a set of power tools that help you work with more focus, speed, and care—without replacing your judgment, wisdom, or experience.

We are becoming super-empowered, as Reid Hoffman and Greg Beato discuss in their book *Superagency: What Could Possibly Go Right with Our AI Future*. We can harness information like never before. I continue to experiment with AI in many ways: whether using it to learn the foundations of quantum computing or as a guide to install Apple CarPlay on my ten-year-old SUV. I have instant knowledge at my fingertips customized to my own level of learning. If I do not understand something it says, I simply ask the AI to reword it in a way I can understand it. AI is supercharging my ability to learn and grow. If AI can make me a more super-empowered individual, imagine what it can do for leaders on their own growth journey, or for organizations empowering their high-performing—or low-performing—employees.

When AI can synthesize large quantities of information, it provides knowledge that has been hiding in plain sight. When AI provides access to knowledge to billions of people who have been denied a formal education, it empowers the disenfranchised. What if AI could tap into the depths of our memories, our values, and our emotions to help us harness who we are and who we want to be?

Don't Get Left Behind

Ever since ChatGPT was released, I have asked leaders how they are using generative AI in their communication. They often demur, saying the tech department is working on it. In a highly regulated industry like financial services or a bureaucratic sector like government, these tools are often prohibited. This book is not about enterprise-level use of AI; this book is about how an individual leader uses AI to become a better leader. It is up to each of us to learn about these tools so that we can begin to understand how they will affect our organizations, our governments, and our society going forward.

Before We Begin: Navigating AI with Eyes Wide Open

You may have a few questions in your mind as you go through the book. Here are some disclaimers to keep in mind.

YOU NEED AN AI POLICY

For all the ideas I present, consult your organization's AI policy before using these tools at work. If your organization does not have an AI policy, now is the time to create one so your employees know the boundaries of safe and acceptable AI use. Do not input personal details such as Social Security numbers or medical records unless you know that the AI tool has been built in a way that safeguards that information. Most of my clients work in large institutions. I have found the larger the institution, the more risk averse it is when incorporating new technology. While these institutions are still using AI, they are adopting it with greater caution. For example, when I keynoted for the Institute of Internal Auditors in 2025, all the exhibitors were promoting AI platforms that balanced efficiency with accountability, recognizing the very real constraints internal auditors face in their work.

AI CAN HALLUCINATE OR MAKE UP INFORMATION

Generative AI works by predicting the next likely word in a sentence. It does not *know* or *understand* what it is saying, it is simply generating the most likely response based on its training data. As a result, it can fabricate numbers, facts, or names while confidently asserting that it is right. We need to treat anything AI suggests as an "unverified source" and fact-check it. AI experts commonly recommend using AI in areas where you already have a strong subject-matter knowledge, so that you can more easily spot errors.

AI CAN DEMONSTRATE BIAS

Large language models are trained on a number of different sources, including the Internet. While we lack clarity in exactly how those models were trained, we do know they include information from the Internet that could be biased. If the underlying information in the data is based on biased information, then we have to consider how that bias will show up in the output. Companies building these models are working to reduce the bias, whether for ethical reasons or business reasons, and more needs to be done to understand the inner workings of these models before we rely on them for decisions that affect people's lives. There are also questions about the use of copyrighted data in how the models were trained. We have not yet seen the resolution of this issue and the implications for content creators upon whose content these models were trained.

THIS BOOK IS FOR ANY LANGUAGE MODEL

This book is agnostic in terms of which of the foundational language models you use, such as ChatGPT (OpenAI), Claude (Anthropic), Gemini (Google), Copilot (Microsoft), and Grok (X). Tomorrow, there will be other tools. The underlying strategies in this book apply regardless of which tools you use. When I use the phrase "AI tool," it could refer to a language model like one mentioned above, an app, a website, or any other way in which we have access to AI, now and in the future. New tools will continue to emerge, but the core question will not change: *How do we use AI to be better versions of ourselves?*

This book focuses on mindset and strategies. I am writing this book in the spring and summer of 2025 and will discuss the AI tools that are commercially available today and accessible by someone like me: a senior-level business professional with no technical experience. By the time you read this book, the technology will have changed, but the questions, ethics, and use cases will remain relevant.

TO THE SKEPTICS: WHY YOU'RE RIGHT TO WORRY

I want to speak to those who are feeling unsure, or uneasy, about the idea of using AI to augment their authenticity. If you find yourself thinking, *Isn't this dangerous? Are we really handing over so much power to machines?*

You are right to be doubtful. This book is not an argument that technology makes us better by default. And it is not an argument to simply hand over your authenticity to AI. One of my clients, during a recent executive webinar, said the following: "Allison, you know me. I'm all about authenticity. I don't even want to download ChatGPT because I don't want it to ruin my authenticity."

This book is an exploration of how we can remain human—perhaps even become more human—in a world where machines are increasingly fluent in the language of leadership. AI should not replace your judgment, intuition, or integrity. It should not outsource your leadership. I believe AI can reflect your values back to you, challenge your thinking, and serve as a patient, judgment-free thought partner. Rather than shield yourself from AI, learn how to use it effectively, learn where the boundaries are, and learn how you can become a better leader as a result.

AI is not an answer—it can serve as a mirror. And like any mirror, it is only helpful if you take the time to look into it and think about what you see. I encourage you to be skeptical and to question what you read here and elsewhere. Your questions are not only valid, they are also essential. Bring them with you as, together, we explore how this technology can help you lead more authentically, not less. Let us proceed with our eyes open, our values intact, and our voices clear.

At its core, leadership is a deeply human endeavor. Let us start with what connects us as humans.

PART I

The AI Authenticity Framework

CHAPTER ONE

Maintaining Authenticity in an AI World

In my work around the world for over twenty years, I keep coming back to authenticity as a source of strength for leaders. Authenticity can refer to the origin of an object or a piece of music. We often refer to authenticity as being "real" or "genuine," but who decides what is real and genuine? Fascinating research suggests that authenticity is something we bestow on others when we see them acting in a way that validates our expectations of how they should act.

When I advise leaders on how to tap into their authenticity, I define the concept as *speaking and acting in alignment with your values and beliefs*. Whatever actions you take or words you choose, you remain true to your personal or professional values. You aim to consistently show up as your best self.

Authenticity is risky and requires moral courage. It is about having the integrity to speak the truth, even when it is uncomfortable, and the discipline to lead with your values when it is easier to just go along with others. It is not always safe to be authentic when you can be politically or physically neutralized due to your beliefs. Leaders often struggle with this concept when their beliefs go against the grain of their organization or their industry.

I do not promote authenticity at all costs. In my keynotes, I discuss the importance of *situational authenticity*, which means we can show up differently in different situations. For instance, at home I might wear jeans and sandals, while on a keynote stage I wear a dress and high heels or sparkly sneakers. In doing so, I am choosing situational authenticity, finding a balance between what feels right and what is effective for my audience.

Authenticity Is Your Strategic Advantage

The way in which you communicate either builds or erodes trust, both in you and in your organization. At a time when trust in leaders and institutions is on shaky ground, authenticity is your strategic advantage. It is one of the critical ways in which you earn the trust of your leadership team, your shareholders, your clients, and the general public.

When an audience sees you speaking and acting in a way that is *not* aligned with what they know about who you are, they mistrust you. They wonder who you *really* are and what you *really* think, and they put up a protective barrier. This creates an environment of pushback, passive aggressiveness, and dissent, which can ruin an organization's culture. It creates organizational drag, stretches out decisions, and bogs down implementation. It can cause a reduction in employee satisfaction and increased turnover. In short: A lack of authenticity can bring down your organization.

The good news is you have that authenticity within you. We all do. We do not need to create it; we need to simply find it within us and then find the best ways to let it out.

Inside-Out Leadership Is Not Easy

This is not the traditional top-down hierarchical model of leadership. Instead, it is *inside-out leadership:* By sharing with you what drives me in my leadership, I inspire you to share your own personal motivations with me and with others, so that together we can achieve a shared vision.

This is not easy work. It takes courage and vulnerability. When leaders ask me, "What is the biggest mistake people make when they communicate?" my response is, "You do not let yourself out. You show up as the leader you think people expect, instead of the one they actually need. You stick to the corporate talking points and stay within a safe, corporate box." What does this sound like? A flat voice reading from a script that was written by someone else. Going through the motions. Checking the box that you "delivered the message." As a result, people cannot see you, which prevents them from relating to you, which holds them back from trusting you.

How AI Encourages Faceless Leadership

AI can encourage this facelessness, because it provides the next likely word in a sentence. The next likely word is based on all the words in the AI system's data set, which is, by definition, inauthentic. AI can promote colorless, bureaucratic jargon over human language. For those leaders who do not care about authenticity, AI becomes an easy way to further remove themselves from the people they serve, such as by sending their AI avatar to a meeting or asking the AI to fire an entire department. When they outsource their leadership, they strip themselves of their effectiveness.

If we define authenticity as "true to the original," then the more we let AI take over our communications, the more humans will doubt our authenticity. And when someone doubts your authenticity, you lose their trust. Use AI to bring out your best self as opposed to outsourcing your voice to artificial intelligence.

After seeing the damage AI can do to a leader's authenticity, I identified boundaries and best practices for when and how to use AI in the process of communicating a message. I developed the AI Authenticity Loop as a framework to guide leaders in using AI every day in a way that enhances as opposed to outsources their authenticity. The term "loop" comes from the concept of "keeping the human in the loop," a critical component of AI development and use. Those developing AI systems will use this term to refer to the importance of human input, human feedback, and human oversight when training these systems to ensure they perform in the way we intend. Do a Google search for the "Paperclip Maximizer" for a theoretical example of the unintended consequences of AI acting without human oversight.

Why Humans Still Win

The first time I attended the MIT EmTech Digital conference in 2022, which provided a crash course in AI development, training, and deployment, I left thinking that AI could never do what I do for my clients. It simply did not have the language capabilities and the context and nuance of human understanding that is required when we communicate high-stakes messages. By November of that year, OpenAI had released ChatGPT, and at the 2023 MIT EmTech Digital conference, AI's capabilities had exponentially improved. I still believe AI will not replace my work. However, it provides a growing number of opportunities to *augment* what I do with clients.

I continuously compare the AI use cases I describe in this book with the same work I personally offer to my executive clients. Who does it better: a human with lived experience or an AI trained on human knowledge? The reason is partly to test the capabilities of the tools, and partly to test my job security. In the field of authentic communication, *humans still win over AI*. I believe that the quality of feedback and coaching of human experts with lived experience is far superior to what the AI provides in most areas—with the exception of synthesizing large quantities of information. While my own perspective is certainly subjective, it is nevertheless important to share that when I experiment with these AI tools to replace something I would have done for clients, I repeatedly see how humans do it better. It is for this reason that I believe AI can augment what humans can do without replacing them. Technology will evolve and these AI tools will improve; however, I still believe that humans will value and benefit from the input of other humans.

My goal is not for my clients to use AI instead of working with me. For the types of situations in which they need support—high-stakes communications that require key moments of leadership—they want a trusted human with lived experience. I would like my clients to realize they have *additional* support available that can supercharge our work together. And for those who lack access to a human advisor, AI can help in ways they would have never been able to afford.

The use cases I share later in this book are not a comprehensive list of ways AI can make us more authentic. There will be more, and each one has areas where I have tried to use AI and it failed. My goal is to provide a compass for AI use going forward as technology evolves.

CHAPTER TWO

The AI Authenticity Loop

Step 1—It Starts with You

Communication starts with you. You are sitting in your office, at home, or on a plane, thinking about an idea or challenge. You see an opportunity to speak up, influence others, and in doing so, make a positive impact. Capture that moment: If you are a visual learner, write down the challenge. If you learn by listening, speak it out loud to yourself.

You might be asking:

- How can we address this critical business problem?
- How can I convince leadership to increase our budget?
- How can I gain buy-in from my team on this new initiative?
- How do I ensure I don't fall on my face when delivering opening remarks on behalf of my company at a community event next week?

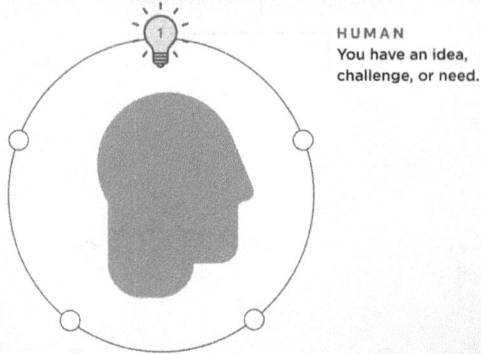

Figure 2.1

We often spiral out of control ruminating on these topics. Instead of getting caught in analysis paralysis, use AI to help you take the next step.

Step 2—Brainstorm with AI

Once you identify an idea or challenge, AI becomes your thought partner. You can ask an AI tool for different ways of looking at the issue or framing your argument, or to conduct initial research on the issue.

Of course, you can consult the humans in your life: your team, colleagues, friends, or family, as well as a trained expert such as myself. However, the AI is instantly available and is judgment free, unlike some of your friends and family members.

When I had a larger team, I would waste precious time trying to figure out who on my team I should ask for guidance and what method I should use to ask them, and not enough time actually

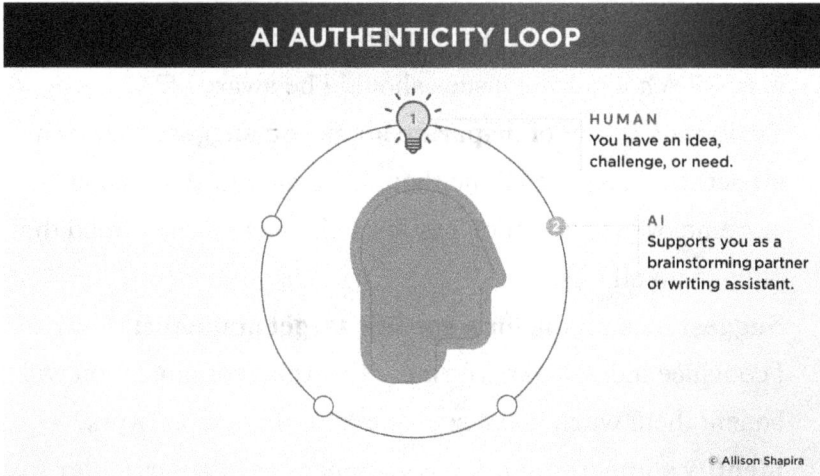

Figure 2.2

addressing the issue. AI can augment the human support you already have or provide assistance to those without that human support.

AI's Common Uses (Before We Go Deeper)

At this stage of the writing process, there are a number of ways people are using AI—some obvious and some non-obvious. Let me briefly touch on some of the most common ways people use AI at this stage, before diving into the ways AI can bring out your authenticity. You can ask AI to:

- **Brainstorm ways to address an issue:** What is a unique way to make this budget request urgent and timely?
- **Research a topic or an audience:** How is the supply chain industry affected by recent tariffs?
- **Summarize large quantities of information:** What are the

key themes from the latest employee or client satisfaction survey? What glaring issues should I be aware of?

- **Evaluate a group of disparate ideas and suggest a speech structure:** Here are all the thoughts I want to share about our employee satisfaction results; help me find the thread that pulls them all together.

- **Suggest messaging for a specific target audience:** How can I convince individual performers that this reorganization will benefit them when it just seems to give them more work?

- **Identify some unintended consequences of a particular course of action:** How can I make sure I'm not stepping on an emotional landmine when I propose this new policy?

- **Create powerful slides that make a message come alive:** Take this text-heavy presentation and create more visually appealing slides that people can read from the back of the room.

Searching for Boundaries

What is the boundary of acceptable vs. unacceptable AI use? We are defining those boundaries as we speak, and everyone draws the line in a different place. To answer that question, I ask myself, "Would it be OK if another human provided this kind of help?"

- Would it be OK if a colleague helped me brainstorm an idea? Yes.
- Would it be OK if an assistant researched a topic? Yes.
- Would it be OK if a graphic designer created slides for me? Yes.

Would it be OK if someone else wrote the first draft of my presentation? *It depends.* If you are crafting a single message, AI can be incredibly helpful in getting started. Many of us remember the first "aha" moment in which we realized the power of AI. I remember that moment for a colleague of mine, another communication professional. He was struggling with a difficult message to convey to a client and felt paralyzed by how to delicately phrase it.

One day, we were walking down a hallway as he confided his overthinking of this message. We stepped into a nearby conference room, and I opened my laptop. I pulled up ChatGPT and quickly typed in what my colleague wanted to say, adding in the complexity and nuance of the situation that was holding him back. He was floored when, within five seconds, he had a solid first draft. Far from being perfect, it nevertheless gave him an idea of how to phrase his message. It jump-started his creativity in a way that helped him create his own language. It provided suggestions, but, ultimately, he retained cognitive agency over the final message.

When AI Helps vs. when It Hurts

I *do* recommend using AI to overcome a mental block when crafting a message. However, I *do not* recommend using AI to write the first draft of a speech or presentation. Why is that? Through the act of speechwriting, we uncover what we want to say. As we craft a message, we have an inner dialogue in our head about the message, the audience, and the potential ramifications of what we are crafting. That inner dialogue is critical as we create our messaging. In fact, it constitutes part of the strategy behind the message.

When we outsource the first draft to AI, what the AI provides is an external document devoid of that strategic thought process. It may not be the right strategy. Even if it is, it will take longer to edit and make your own, and it will ultimately be harder to remember because it did not come from your brain.

You may disagree. Writing is a very personal process and, as I am finding, AI use is very personal. Each of us finds the tools, processes, and pacing that works for us. However, consider copyright protection. As of May 2025, the US Patent and Trademark Office said that copyright does not apply to AI-generated content, although it might apply "if a human selects or arranges AI-generated material in a sufficiently creative way, or modifies material in a way that meets the standard for authorship." Other countries have different policies. If you are a corporate executive, anything you say represents your organization and becomes part of its intellectual property. If you are a communications professional drafting a message for a client or leader, the same expectation applies. Keep that in mind when you use AI in your communication process so you do not create legal challenges for your organization or your business.

With those common use cases covered, let us now look at three specific use cases for how AI can bring out our authenticity.

CHAPTER THREE

Three Ways AI Can Make You More Authentic

These use cases start at the surface level and then dive into deeper areas of authenticity. Before you read them, think about the situations you encounter at work, in which you have to deliver difficult messages, and consider how you would apply the suggestions that follow to each situation.

I: Break Through Corporate Speak

How many acronyms exist in your industry, or even within your organization? Some companies I advise are so large that they have different acronyms for the same concept across different lines of business. Many of us live inside a corporate box that is full of jargon. Some people use jargon to demonstrate they know what they are talking about. If you really know what you are talking about, you do not need jargon; and when you *do not* know what you are talking about, no amount of jargon will save you.

We are so accustomed to jargon that we do not even realize when we use it. In consulting, you're "crunching on a deliverable." In finance, you're going to "socialize" an idea and then "revert." When you revert, you'll "pull up" to discuss it.

I remember advising a senior leader from UNICEF, helping her fundraise for the organization's critical work in war-torn countries around the world. She kept repeating the phrase, "We are repatriating children associated with armed groups back to their communities." I asked her, "What does that mean?" She replied, "Basically, we're sending child soldiers home to their families." While the first phrasing was what she needed to get her budget approved by the home office, the second phrasing was what inspires individuals to support a cause.

When we use jargon, our words lose their power. As a result, we miss an opportunity to inspire people to take action. What if AI could reduce the jargon in your presentation?

Try this:
1. Input the text of your message into an AI tool, like ChatGPT or Claude or Gemini.
2. Ask the AI to identify potential jargon.
3. Ask it to reword those sentences in more easily understandable language.
4. Ask it to suggest alternative messaging for your target audience, such as senior leaders, top performers, or individual contributors.

Could a human help you with this? Certainly. One of my favorite activities when advising senior executives is to challenge their use of jargon and acronyms, and to help them reword their message so that it is more in alignment with their authentic voice. This is a deeply personal process that requires them to trust me and be vulnerable enough to share how they really feel about a policy or direction.

However, I am not available at midnight before your high-stakes presentation to the board, when you are putting the finishing touches on your presentation. Or perhaps you are in a creative flow and need immediate assistance instead of waiting until the next business day. AI is an instantly available tool to help you maintain your creative focus and energy.

Reducing jargon is an easy step in the process of using AI to be more authentic. We can go deeper.

II: Find Your Voice in the Chaos

You can also use AI to bring out your best self *when you least feel like it.* Authenticity comes in many forms, and it is not always positive. You can be authentically unprepared or authentically long-winded. I am authentically impatient when someone wastes my time, but that does not always make me a more effective leader. It is not whom I aspire to be when I am at my best.

As a leader, while it is critical to allow yourself to feel human emotions, the way you *show* those emotions has a significant impact on the culture of your organization and the productivity (and turnover) of your teams. Leadership is overwhelming. Humans are inefficient, irrational, and inherently imperfect. We do not always act in our best interests.

THE EMPATHY PARADOX: WHEN AI OUT-EMPATHIZES DOCTORS

What if AI could bring out our best selves, on demand? What if AI could remind us of how we *aspire* to act? A fascinating study showed how empathetic AI can be. Published in *JAMA Internal Medicine*, it evaluated physician responses to patient questions on social media and compared them to the same responses generated by an AI chatbot.

The study found that "Chatbot responses were preferred over physician responses and rated significantly higher for both quality and empathy." The AI responses were rated as more empathetic than the doctors' responses. Why is that? Because doctors are busy. They do not always have time to listen. Like all of us, they go on automatic. It is not necessarily their fault; it is a product of mismatched incentives and overwhelming demands in the ecosystem in which they operate. Furthermore, some individuals have different cognitive styles, or neurodivergence, which do not demonstrate empathy in a typical way. The AI did not have those considerations. The AI was trained to be empathetic. It had endless patience and did not mind adding additional words, suggestions, and context to comfort the patient.

In a more recent study performed in 2025, a team from the University of Geneva and the University of Bern found that a number of large language models "outperformed humans on five standard emotional intelligence tests, achieving an average accuracy of eighty-one percent, compared to the fifty-six percent human average reported in the original validation studies." You might ask if the language models performed better than humans because they contained all the answers from those studies in their training data. Regardless, it demonstrates that AI tools can generate the "correct answers" in terms of what makes someone more empathetic.

BEFORE YOU HIT "SEND"

In my advisory work, I help leaders take a moment to "pause and breathe" before they react to a situation or answer an unexpected question. When you are in a leadership role, every word you say has an impact—either good or bad. Your moment of frustration can derail your entire team or affect the stock price of your company. What if AI could help you pause before hitting "send" and choose the tone you actually meant to convey? What if AI could prompt you to be a better human, across cultures and across contexts?

Imagine you spent most of your career on Wall Street, where communication is often direct and blunt. Now, you are giving feedback to a colleague at a global nonprofit who comes from a high-context culture, where communication relies more on tone, context, and relationships. How might you adapt your words to ensure your message is both respectful and effective? Or, how do you provide feedback in a hierarchical political structure where nuance is essential? This is situational authenticity in action. AI can help you not only be your best self but translate your best self across cultures.

Try this:
1. Draft a difficult message you have to communicate to a team member or client.
2. Input it into an AI tool.

3. Ask the AI, "What tone, values, and priorities are showing up here?"

4. Based on the response, ask yourself, "Does this match how I want to lead?"

5. Tell the AI about your personal or professional values and ask it to make your text more empathetic and more in line with your values and the culture of the person or audience you are speaking to.

6. Role-play with the AI out loud and ask it for feedback.

7. Bonus: If you know the personality styles of your team or client, ask the AI to help you communicate in a way that would resonate with their Myers-Briggs Type Indicator, DiSC profile, or Kolbe assessment.

Certainly, there are very capable human advisors who can more effectively help you with this process. They may be on your team or in external agencies, and their expertise is critical. However, small business owners often lack access to this kind of knowledge, so AI democratizes the availability of feedback in a way that is "good enough" to point you in the right direction.

III: Unlocking Your Hidden Stories

When I advise leaders, I teach them to bring out their authenticity by tapping into their core values and remembering the experiences that shaped who they are. I have seen leaders transform when they finally align their company's message with their own mission.

Earlier in this book, I mentioned that when you share your own personal motivation around why you do what you do, it allows others to connect with you on a much deeper level that builds trust. This trust makes you a more effective leader, whether you are running for office or promoting a new corporate strategy. But we often bury this motivation deep within us, after years in a corporate or bureaucratic environment. Or we have internalized a belief that to do so would be unprofessional. This belief holds us back from our power as leaders.

There is one core question that unlocks your authenticity every time you communicate, even in the most professional of settings. Since introducing it in my bestselling book *Speak with Impact* in 2018, it has become the cornerstone of my methodology.

WHY YOU?

Before every meeting, pitch, or presentation, ask yourself: *Why you?* That question does not mean *How many years of experience do you have?* or *Why are you the best person to speak about this topic?* That question really means, *Why do you care . . . about the topic you're addressing, about the people you represent, or about the impact you want to have? And when was a moment in your life that made you care?*

Usually, people's first answer to "Why you?" is superficial. They share the socially acceptable answer. When I advise leaders on how to build their confidence, I frequently need to push them to dig deeper into their answers to this question, to ask "Why" more and more, so that they arrive at the core message they want to convey.

Your "Why you?" statement could be because you are passionate about solving problems for others, or you believe in serving a greater good, or you want to ensure the future prosperity of your country. The answer usually reveals powerful motivators that make you who you are, and revealing those motivators to others helps you build trust and connection. When you build trust and connection, you are better able to inspire people to take action.

This often lies dormant within us. Have you ever heard the phrase "That person has forgotten more than I will ever know"? We use it to refer to someone with an incredible breadth of wisdom and lived experience. Each one of us has unique perspectives. There are formative experiences that have shaped who we are, relationships that taught us our core values, and seemingly mundane anecdotes that led to lifelong lessons.

YOUR PERSONAL AI SPEECHWRITER

What if you could train an AI tool on your values and beliefs, your stories and experiences, and your tone and style? You could upload articles you have written or speeches you have given. You could input your "Why you?" statement and have the AI patiently ask you questions, one at a time, to call to mind your most formative experiences to arrive at what truly drives you, in your work or in your life. Then, you could ask that AI to draft a message or outline a presentation in your style, based on your stories.

Or, if you operate in a culture where showing emotional vulnerability is not widely accepted, you could use this AI tool to privately discuss emotions and ideas that you could not share publicly. It becomes a first step to being vulnerable: sharing with an AI before you share with another human. AI becomes a nonjudgmental coach that gives leaders a place to practice expressing difficult emotions, experiment with showing empathy, and explore the deeper "Why you?" stories without losing face. What if you could input your organization's corporate talking points into that AI and then ask it to reword the corporate messaging *in your voice,* finding relevant stories from your past to reinforce it? Imagine how much more effective you would be when speaking at a community event or at an all hands meeting.

THE PRIVACY PRICE OF AI INTIMACY

How is this task any different than the job of a speechwriter? It is different in a few ways, some positive and some negative. First, the speechwriter is limited by the amount of time they can spend with you and the number of questions they will ask. An AI system could review your entire life with you. Second, you build trust with the speechwriter. You trust that person to keep your conversations private, and you believe that they will not feed your stories into some corporate database that could be accessed by others.

What would it take for you to trust a single company with your most intimate stories or life experiences? You already tacitly trust Apple, Samsung, or Google when you send text and email messages, store your credit card number, or use your computer as a personal journal. What would it take to trust them with a database of your memories?

Like many AI tools, this use case will not necessarily disrupt the role that a speechwriter would play, it will simply make these capabilities more accessible to those who could never have afforded a speechwriter in the first place. And for those who *can* afford a speechwriter, their high-stakes presentations will always benefit from expert human feedback.

To the speechwriters: Imagine creating a custom chatbot for each of the leaders you support, helping you keep track of each one's stories, anecdotes, and style to help you generate ideas. It becomes an organized, searchable database. The leader still relies on your judgment and experience; you simply have more efficient tools to harness and remember that leader's style. When we merge an AI that contains a synthesis of human knowledge with the targeted wisdom of an expert's own lived experience, we create a powerful combination.

This does not just apply to speeches, it applies to everyday conversations for which you are struggling to find the best way to connect with someone and build trust. It helps you search your own database of thoughts to call forth the best way to respond to a difficult situation, such as a coaching conversation with a low-performing employee or a confrontation with a colleague.

This is not the future; you can do this right now. The technology is getting better and better. If we compare the current state of generative AI to the Internet, you might say we are somewhere between dial-up and DSL in its level of development.

When Students Can't Stop Talking to AI

For over ten years, I have taught a graduate course at the Harvard Kennedy School called The Arts of Communication.

In 2024, Harvard faculty members Teddy Svoronos and Sharad Goel, together with developers Evangelos Kassos and Joe Nudell, created an AI tool called PingPong where faculty could create chatbots for their own courses, similar to a custom GPT from OpenAI.

For my course, I created a series of custom chatbots, including one called "Finding my Authentic Voice." I wrote an elaborate prompt that included my methodology, based on the question, "Why you?"

One day, I was delivering a workshop about AI and authenticity at the Harvard Kennedy School. During the workshop, I asked each student to interact with a scaled-down version of that authenticity chatbot. They sat silently next to each other for two minutes: It was the only time I ever enjoyed seeing all my students on their phones. Then, I asked them to discuss that experience with the person next to them. They ignored me and kept typing with the bot. Finally, I said, "Please stop chatting with the AI, and talk to the human next to you." Everyone laughed, put down their phones, and turned to their neighbor.

Afterward, as we debriefed the Authenticity GPT experience, the responses were startling. I asked, "What was it like answering the chatbot's questions?" Students described the experience as both spooky and revelatory. A man in the room said, "I remembered a story from my life that I hadn't thought about in twenty years." A woman in the room said, "The chatbot asked me a question that made me feel seen."

What does it mean that, after two minutes of chatting with an AI, we find it hard to tear ourselves away to talk to a human? What does it mean for a nonhuman entity to guide you in finding your authentic voice? Right now, we are conducting these social experiments at a global scale, and I believe we need to do more research in order to understand the impact of this use case on human relationships and intimacy.

This is not new; in 1966, MIT professor Joseph Weizenbaum created a program called ELIZA. It worked using simple pattern-matching rules to reflect a user's statements back to them as questions—mimicking the style of a nondirective therapist. What surprised Weizenbaum was not the technology, but the way people emotionally bonded with it. He became increasingly alarmed by how easily humans could attribute feeling and understanding to a machine. We will explore the implications of that in the next section.

Try this:

1. Try out my Authenticity GPT at allisonshapira.com.
2. Spend five minutes chatting with the tool. Answer the questions, "Why you? Why do you care about your work, and when was a moment in your life that made you care?"
3. Input a corporate or bureaucratic message into that same AI chat.
4. Ask the AI to make suggestions on how to reword this new message in your voice, based on the values and beliefs it learned about you from your "Why you?" statement.

CHAPTER FOUR

But Is It Really You?

You might be tempted to stop at this point and simply use the responses AI provides you, but you are less than halfway through the writing process. You might also get caught up in the suggestions of Chapter Three, especially if you start reliving your past experiences. I highly recommend you consult another human when doing this, in order to process the experience.

Step 3—Make It Yours

In Step 3, it is your responsibility to interpret the AI's suggestions and use critical thinking to evaluate which components to use.

Some will simply copy and paste what the AI writes. This is a mistake and will negatively affect your authenticity. It could also be unethical. There are numerous examples of lawyers citing fake court cases in their court filings, because they used AI without fact-checking what it generated.

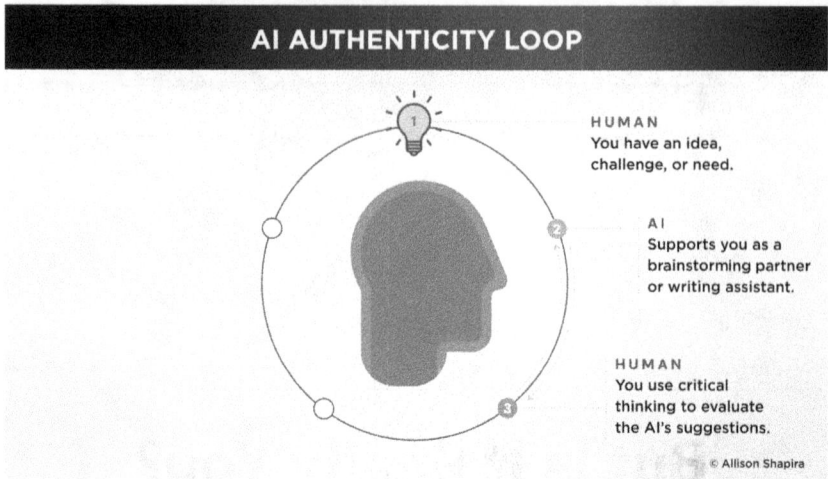

AI AUTHENTICITY LOOP

HUMAN
You have an idea,
challenge, or need.

AI
Supports you as a
brainstorming partner
or writing assistant.

HUMAN
You use critical
thinking to evaluate
the AI's suggestions.

© Allison Shapira

Figure 4.1

WHEN AI MAKES YOU INVISIBLE

Copying and pasting can also be obvious to the listener or reader. For example, I recently hired a new executive assistant. During the search process, I received dozens of resumes and cover letters. Ninety-five percent of those cover letters looked like they had been copied and pasted from an AI tool with the instructions *Look at this job description and my resume and come up with a cover letter that incorporates language from the job description to describe my experience and why I'm the right person for this job.*

I understand how the maddening ordeal of looking for a job will make that process tempting—however, it was ineffective for me as the hiring manager. I interviewed a number of candidates, but ultimately the person I hired was one of the few candidates whose cover letter was written in her own voice. It was not just because of the cover letter—her qualifications, references, and interests made her the right fit—but because of the cover letter, she stood out from the crowd.

When you are evaluating what an AI comes back with, my colleague and fellow speaker Terry Brock says "copy, paste, customize." Ask yourself: *Is this accurate?* Remember, generative AI is a predictive tool, so you will need to fact-check any specifics it provides, such as names, numbers, dates, or concepts. Ask yourself: *Is this authentic? Does this represent my views? Does it pass the "billboard test": Would I feel comfortable if this message were on a highway billboard in my hometown, along with my name and photo?*

You may go back and forth between Steps 2 and 3. Brainstorm with AI, then review with a critical eye. Consult other humans for their guidance as well. Repeat until you feel like you have arrived at what you truly want to say. You are responsible for the message you deliver in any format. And since your ultimate goal is to build trust, ask yourself if the language you are using will build or undermine the trust of your audience.

Step 4—Practice with AI

The difference between a good communicator and an inspirational leader is the way in which they hone their craft.

The inspirational leader does not just deliver talking points; they practice their message and seek external feedback to continuously improve. They do not rely on the words themselves to deliver the message; they ensure every part of them—including their voice, body language, and tone—communicates the same message. Once you have written your speech, presentation, or message, AI can help you by providing feedback.

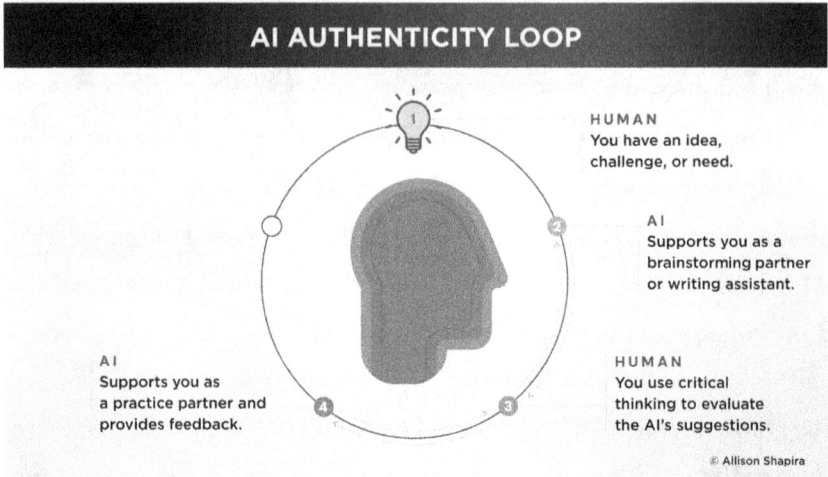

Figure 4.2

POLISH YOUR MESSAGE

Here are some of the ways people will use AI for feedback on their written message:

- **On behalf of your target audience.** How would a particular client persona or personality profile respond to this message?
- **On behalf of your worst critic.** You know who the naysayers are in your organization. How would they respond to this and what can you do to avoid or prepare for their pushback?
- **On the structure of your message.** Does your presentation follow a logical flow? Where can you add better transitions?
- **On your grammar.** Where can you adjust the language to make it either more accurate or more formal? Or less formal?
- **On your clarity.** Where are you verbose, and how can you tighten the language to achieve greater clarity of message?

What Would Allison Say?

For years, I have struggled to create an effective rubric to use when evaluating a speech or presentation. Having worked in this field for over twenty years, I have an elaborate rubric *in my mind* that I activate when someone is delivering a message to me. But how do I share that with others in a way that is effective? The more extensive the rubric, the harder it is to use, but the more helpful the feedback.

An elaborate rubric is too complicated for another human—but perfect for an AI. So, this year, I tried something different for my students at Harvard. Using the newly available PingPong tool described on page 24, I created a custom AI chatbot called "What Would Allison Say?" To create this chatbot, I wrote a multi-page prompt that included everything I look for when evaluating a written speech, such as persuasive tools, uniqueness, messaging, structure, and use of personal story. I included examples of strong speech openers and closers, and built out what I believe effective speeches look like. The result was a chatbot that provided feedback on a written message, based on what I look for when I advise leaders. When students are preparing for their speeches, at any hour of the day, they could add their text to "What Would Allison Say?" and receive my guidance.

You might ask, could they ask me these questions during office hours? Certainly. But office hours last only twenty minutes and take place just once a week. "What Would Allison Say?" is available on demand. This chatbot also helps students who were not able to take my class. So often students will come to me for guidance on their commencement speech or job interview. In order to protect the time I have available to everyone, I can now say, "First use my chatbot, incorporate the feedback it gives you, and then bring me your second draft." This ensures I remain available, but at a more advanced stage of the process.

Allison in Your Pocket

Currently, I am developing an AI tool to share with my executive clients as well. For years, my clients have been telling me that they have Allison "sitting on their shoulder, whispering in their ear" when they present. Soon they will have "Pocket Allison" available to help them twenty-four hours a day, seven days a week, with the option to personalize it based on notes from our sessions. They still have access to me during the length of our engagement, but they can ask my chatbot and perhaps even my avatar questions any time they want, and have the benefit of my experience, on demand, customized to a particular situation.

When Humans Are Busy

I have experimented with this myself. My own business coach encourages me to record our coaching sessions so I can review them in the future. But reviewing meeting notes is time-consuming and inefficient. Instead, I took the transcripts from all of our sessions, removed any confidential details, and created my own personal chatbot based on that coach. Whenever I struggle with a client proposal or strategic business decision outside of our normal coaching hours, I can simply chat with this bot and reality-test ideas instantly. I can also email my coach—but then I have to wait for a reply. Even a ten-minute wait takes me out of my flow, so this AI tool provides a "good enough" response instantly to keep me focused.

Here is another example from my business. When I was writing my keynote speech, *AI for the Authentic Leader*, I wanted the guidance of my chief operating officer. I knew she was busy with other tasks, and I would want to give her time to fit this request into her schedule. However, I wanted feedback right away. So, rather than ask her to prioritize my keynote over our company's strategic priorities, I input the text of my keynote into Claude, and within ten seconds I had substantive feedback on where to improve the speech. I still asked my COO for feedback, but I also had the benefit of immediate AI guidance. These tools should not replace human judgment—they provide additional support in between your meetings with those humans.

THE ENERGY YOU PROJECT

Our message extends beyond the words we say and includes the energy we project. As a leader, the moment you enter a room, your energy affects the entire room. You are communicating before you even say a word. It takes many forms, including the way you carry yourself, your facial expressions, your pacing, and your tone of voice. This relational energy is contagious; you determine whether people feel excited or anxious. How intentional are you about the energy you bring into the room or on a video call?

In my two-day executive workshops, the power of energy is often a major revelation for the participants. Until they observe the impact of a speaker's energy on how effectively that speaker delivers a message, they do not understand how impactful it is. Outside of that learning environment, it is incredibly challenging for us to evaluate this ourselves, so AI can provide feedback on our energy.

Practice without the Pressure

One of my favorite AI tools for this purpose is Yoodli.ai. Years ago, when app developers started coming to me with their AI tools, Yoodli stood out. You can either upload or record a video of yourself speaking, and Yoodli's AI tool will provide analytics on your pacing, pausing, and use of filler words. Since then, Yoodli has extended far beyond pacing, filler words, and the mechanics of speech to include content structure, delivery, and the ability to customize a platform around any rubric. Yoodli has built out an elaborate series of role-play scenarios, such as sales calls or job interviews, with AI avatars. In addition to being one of Yoodli's early adopters, I also created a bespoke version of the platform, customizing the AI to my own books and videos. It is an incredibly helpful practice tool for my clients.

When I first met Yoodli's co-founder, Varun Puri, what struck me most was that the platform also integrated humans into the feedback process. You can share your recording with a friend, coach, or advisor, who can then provide time-stamped feedback. It is the combination of human and AI feedback that I find most powerful and most effective. My advisory clients can record their presentation and send me a link to view it, and I can provide feedback in the moment or at my convenience.

It is not easy; one of the biggest barriers to using tools like this is that no one wants to see themselves on camera. In fact, during executive workshops, my team and I will record leaders and play back the video, right there in the room, in front of the other leaders. Participants consistently say it is both the *hardest* activity and the *most effective* activity, because it forces them to see how they come across. While it is a painful experience, I believe it is far better to cringe when watching yourself within a safe learning environment, as opposed to watching the video of your actual performance in front of 500 colleagues or clients. Yoodli provides a safe, judgement-free learning environment.

Try this:
1. Open a free account at Yoodli.ai.
2. Record a video of yourself speaking for at least one minute.
3. Review the AI-generated feedback.
4. Share your speech with a friend, colleague, mentor, or coach for their feedback.
5. Before a difficult conversation, role-play that conversation to ensure you use the most effective messaging.

THE PARADOX OF FILLER WORDS

For years, leaders have attempted to reduce the filler words they use, including *um, ah, like,* and *so.* Every language in the world has them, and we often pick them up from others, in what Professor Tim Murphey of Kanda University of International Studies calls "linguistic contagion." Used excessively, these vocal disfluencies can distract an audience and make the speaker look unsure of themselves. If every sentence is punctured with an *I mean* or *kind of* or *just*, it reduces the power and punch of your words. Imagine saying, *I just kind of think that this is important, right?* instead of *This is important.*

There is nothing wrong with a few filler words, and they can even play a strategic role, as I discuss in my article for *Harvard Business Review*, "When Filler Words Are Actually Useful." In addition, I will never forget the critical feedback I received one year from my student evaluations: *For all the emphasis you place on authenticity, your focus on reducing filler words feels overly perfectionist.* While that hurt to hear, it was a wake-up call for my teaching methodology.

In fact, one AI tool demonstrates the importance of these tiny words. Google's NotebookLM includes an Audio Overview feature that generates two incredibly lifelike AI voices discussing any document(s) you upload to the platform, creating an on-demand podcast. In my experience, NotebookLM is often responsible for a person's first mind-bending AI experience, where AI challenges what they thought was possible. Last year, I showed this tool to my eighty-one-year-old mother and created a podcast about her. I'll never forget how her jaw dropped open as the AI voices brought to life significant moments from her life.

Recently, I listened to a podcast interview with Raiza Martin and Steven Johnson from Google Labs, about how they developed NotebookLM's Audio Overview feature. What they found was fascinating. They needed to add in filler words like *um* and *ah* and backchannel responses such as *uh-huh* and *mm-hmm* so that the voices were more relatable.

What a paradox: *Humans are using AI to reduce their filler words, while AIs are using filler words to sound more human.* This shows the delicate balance we strive for in situational authenticity. I sometimes find myself allowing grammatical errors to show the language comes from me and not an AI. As my friend and colleague Matt Strain, AI advisor and founder of The Prompt, said to me, "As AI-generated writing becomes more pervasive, some writers are starting to leave subtle signals that their work is human-made. Much like the humility flaw in a Persian carpet, we're now seeing people leave in an intentional typo or imperfection—a quiet rebellion against the too-polished edge of machine-generated perfection." I myself wonder if these "artifacts of humanity" will soon lose their utility as AIs strategically insert their own flaws to appear human.

Step 5—Own Your Message

We have arrived at the final step. At this point, you have refined your message, practiced your delivery, and reviewed your feedback. The last step of the process is all about you, not your avatar or hologram (at least not yet). You feel your heart beating loudly in your chest as you try to calm your nerves. You take deep breaths to center yourself. You remind yourself why your message is

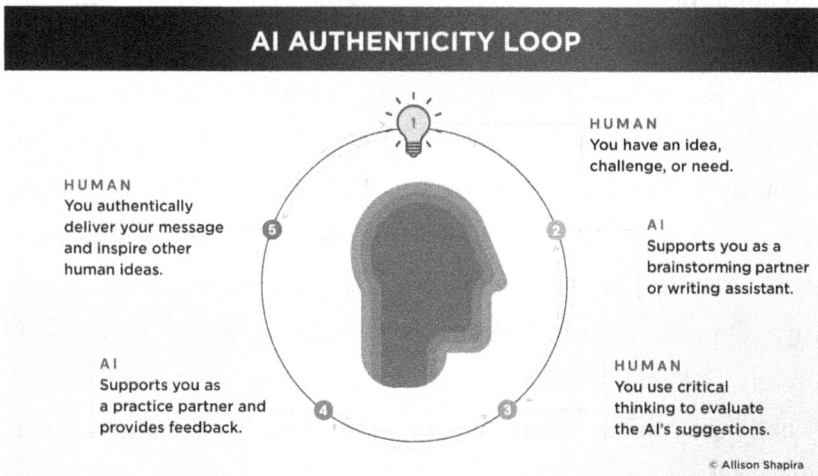

AI AUTHENTICITY LOOP

HUMAN
You have an idea, challenge, or need.

AI
Supports you as a brainstorming partner or writing assistant.

HUMAN
You use critical thinking to evaluate the AI's suggestions.

AI
Supports you as a practice partner and provides feedback.

HUMAN
You authentically deliver your message and inspire other human ideas.

© Allison Shapira

Figure 4.3

important, and why others need to hear it: You are acting in service of the people you lead.

Leadership requires the courage to speak up, and now is the time. You deliver your message: on a stage in front of a crowd of thousands, in a boardroom in front of the leadership team, or on video during an all hands call. You take ownership of your message and your ideas. Your authentic energy and enthusiasm infuse the audience with your passion and purpose and you inspire them to take action. Your audience receives the message, feels the collective effervescence, and is inspired to come up with their own ideas. And the process starts all over again. Communication starts and ends with humans.

The Human Is the Loop

This is the AI Authenticity Loop: a circle of human and AI collaboration. You can use AI to brainstorm and practice the message, but you always filter it through the lens of your authenticity, expertise, and humanity. *You are not just keeping the human in the loop—the human* is *the loop.*

Use the AI Authenticity Loop to craft and communicate any message. Start using it daily to see what it can do; the more you use it, the better it will work for you. In my research for this book, as well as in conversation with friends and colleagues, I am continuously inspired by how people are using AI in new and creative ways. One of the best ways to learn how to use AI is to ask it directly. Set aside an unrushed hour to open up an AI tool, type in a challenge you are facing, and ask AI how an AI tool can help you address it.

When I advise leaders, my goal is not for them to sound like anyone else. My goal is to help them remove the layers that cover up their authentic voice, which they have accumulated over the course of their professional careers. I am prompting them to be more courageous leaders, which means *having the courage to be themselves.* AI can make you more authentic, because it can counteract the forces in your life that are covering up your authenticity. It can serve as a gentle, patient reminder of who you are—at your best—and prompt you to show it, live it, breathe it, consistently.

CHAPTER FIVE

How I Used My Own Method

In 2024, I made one of the hardest decisions of my life. Nearly ten years after I started to expand my company into a global team of trainers and coaches, I realized my passion did not lie in growth. I needed to simplify. The larger we grew, the more unhappy I became managing the company. I was spending all my creative energy on finding new clients or distinguishing us from other training companies. It was not what I wanted to do, and it was not the best use of my passion and skills. It felt like a physical weight holding me back, like trying to run with a parachute tied behind you. And it was impacting my leadership: I observed how my own frustrations were affecting the way I interacted with my team.

What I really wanted to do was focus on writing, executive advisory, and keynote speaking. My passion is teaching and advising others, either on stage or one-on-one. I saw the world changing with increasing speed, complexity, and uncertainty. I did not want to spend my time worrying about payroll when I could spend that time guiding my clients through this pivotal time of technological, political, and societal change.

I still remember feeling the pit in my stomach as I made the painful decision to downsize the team. This was an incredibly talented team of business professionals and experts, whom I had brought on over the years. My clients loved them, and they in turn loved working together. One by one, I had to have very difficult conversations and break up the team. I had to show up as my best self, and I could not outsource this conversation to anyone else. But AI could help.

Step 1: I made the decision to downsize my company and had to communicate this message to each individual team member. I drafted different versions of the message for different members of my team, based on their personalities and personal drivers.

Step 2: I used a confidential, custom AI chatbot I had built to serve as a co-CEO to my business, based on a template created by Paul Roetzer, founder and CEO of Marketing AI Institute and SmarterX. I customized the AI with my company values, current staffing, client personas, and core services, among other details. I input my different messages into the AI and asked how well they reinforced the values of our company. Without using real names, I described my employees. The AI helped me fine-tune my messaging and the pacing of the conversation so that I could show the maximum level of respect and support for my team, using our company values as guidelines.

Step 3: I reviewed what the AI suggested, further adjusting the language so that it sounded like my voice and ensuring that it was authentic to me. I consulted a few trusted (human) mentors for extra guidance.

Step 4: I chatted out loud with the AI to role-play the difficult conversations. I gave the AI non-private descriptions of the individual team members, such as their personality types and other sensitivities, and asked the AI to role-play different scenarios. The AI was endlessly patient. Every time I was nervous about how it would go, I would take a walk and talk through the conversation with AI as my sounding board.

Step 5: Ultimately, I was the one who delivered the message to each member of my team. It was painful every time. However, I was confident that I was showing up as my best self, and that I had done everything in my power to come across in the best possible way. While I was not perfect in how I handled them, I strongly believe those conversations went much better than they otherwise would have.

Each one of us is faced with difficult conversations in our lives, whether with a spouse, child, boss, or colleague. We are increasingly using AI tools to help us through these conversations. This does not have to outsource our authenticity; it can ensure that the best part of us shows up, when our emotions might otherwise take control and lead us to say something we do not mean and will later regret.

Try this:

1. Identify a difficult message you have to deliver.
2. Go through every stage of the AI Authenticity Loop.
3. If you can, identify a trusted colleague or friend who will be present when you actually deliver the message (who is not on the receiving end of the message).
4. Deliver the message.
5. Ask your trusted colleague or friend for feedback.

PART II

NAVIGATING THE RISKS AND THE FUTURE

CHAPTER SIX

What Every Leader
Should Know

My books have always focused on practical applications. A core component of my role as a speaker and advisor is to educate others. This book started out as no exception: a practical guide to use AI to be more authentic. As it evolved, however, I realized it was not just a manual, it was also part cautionary warning and part strategic insight into how I observe AI working its way into our lives.

Many principles of communication have remained unchanged for thousands of years, which is why we still teach Aristotle's *On Rhetoric* in our classrooms today. Although the modes of communication change, from spoken to written to electronic, the basic principles of inspiring another human being—by touching both their head and their heart—have not changed. However, the effects of AI are uncharted territory. As these artificial entities enter our lives, they will affect the interactions humans have with one another in ways we have never seen at this scale.

If you have come this far in the book, I invite you to join me in this second section, as I explore in small vignettes different reflections I have on the future.

The Dark Side of AI

I am inspired by the potential for AI to help us become better humans, and this book has focused on the benefits of using AI to tap into our authenticity. However, I also have significant concerns about how we use AI. Throughout history, humans have used technology for good and for evil. A common expression is that technology is like fire: It can keep us warm or burn down our house. As more and more people gain access to AI tools, we are seeing examples of both. Let us explore a few areas of concern.

PROTECT YOUR VOICE: WHEN ANYONE CAN SOUND LIKE YOU

Throughout my career, I have focused on the human voice. Our voice is unique to us and recognizable to the people we know and love. Like a familiar scent, a voice from the past can take you back in time and fill you with emotion.

When Your Voice Becomes a Weapon

But our voices are being stolen. Today, someone can recreate your voice using fifteen seconds of audio. In the United States, both sides of the political aisle are guilty of this practice. An AI-generated deepfake of former President Joe Biden's voice told people not to vote in the primaries. A news outlet released a deepfake of Republican Kari Lake, mimicking her voice and image in a satirical video during her Senate campaign. This danger is not just for public figures or celebrities; it can happen to any of us. In Baltimore, a disgruntled employee created a deepfake recording of his boss making racist and antisemitic comments. Scammers will impersonate someone's voice in order to extort money from that person's family.

The Safe-Word Strategy

My family and I created a safe word, so that if one of us ever calls another asking for money, we request the safe word. Recently I had to actually use it. As an experiment, I created my own AI avatar and sent it to my mother. I called her excitedly to hear her reaction. "Wait!" she said over the phone. "What's the safe word?" Mission accomplished. This is a practice I recommend for leadership teams as well as for families.

The Impending Death of Truth

The amount of false information being generated by AI is causing us to question everything we see, read, or hear. Why is this so important? You will need to guard yourself against misinformation. You will need to find trusted sources for yourself and curate trusted sources of information for your teams. As a keynote speaker, I believe in-person gatherings like meetings and conferences will be more important than ever, not just because of the collective effervescence they foster, but because they will be the only way we know if someone really said something.

In a recent podcast interview, historian and philosopher Yuval Noah Harari said, "If democracy is a group of people standing together having a conversation, what happens if suddenly a group of robots joins the circle and starts speaking very loudly, very persuasively, and even very emotionally, and you can't tell the difference between who is a robot and who is a human being? If democracy is a conversation between humans, this is the end of democracy." I believe your voice is your personal property and needs to be protected by law. We also need clear labels of AI-generated or AI-delivered content so that we never confuse an AI with a human.

PROTECT YOUR EMOTIONS: YOUR THOUGHTS ARE NO LONGER PRIVATE

I believe the information your voice conveys should be private. When AI tools can read your emotional state and report it to others, without your consent, it invades your privacy.

The company Hume created EVI, a chatbot that calls itself the world's first AI with emotional intelligence. While you speak with EVI, it analyzes your emotional state and then responds in a way that is regulated to meet you where you are. It shows you the emotions it detects in your voice as well as the emotions it uses in its response. I will admit I enjoyed chatting with EVI, but I am worried about its potential for misuse. Sentiment analysis may be helpful for a dentist trying to make a patient feel comfortable before a root canal. However, what about a dictator evaluating the trustworthiness of a country's citizens?

The Mind Reader's Dilemma

I also believe your thoughts are private. Currently, we are seeing life-changing advancements where AI can convert brain signals into spoken words, returning the power of communication to those who have lost their voice. However, the concerning part of that breakthrough is the ability to read someone else's brain signals in order to understand their private thoughts, without their consent.

Nita Farahany's book *The Battle for Your Brain* opened my eyes to the importance of cognitive liberty, protecting the privacy of our inner thoughts and feelings. Authenticity comes from feeling safe, including the safety that our thoughts and emotions are private to us until we choose to share them. The authentic leader will need that safety for themselves and they will need to create an environment where their teams and colleagues feel safe in the privacy of their own emotions.

PROTECT YOUR AGENCY: WHEN YOU CAN INFLUENCE AT SCALE

Throughout history, leaders have used persuasion and influence to manipulate the masses. We saw it in fiction such as the George Orwell book *1984*, and we saw it in reality with Nazi propaganda during World War II.

Our beliefs are malleable, shaped by our experiences, our communities, and the information we consume. They have been shaped by social media and now they will be shaped by AI. I used to teach my graduate students that no one knows exactly how to persuade every person on Earth; what makes something persuasive to any given person is a complex mix of that person's background, beliefs, and context. "Let's be grateful we don't know how, I would say, because if one person had that information, they would wield a power no one person should ever have."

That reality has changed. When an AI system can access vast amounts of personal data about you, from your buying preferences to your voting habits to your social media activity, it is capable of a level of persuasion we have never seen before. That AI system can generate and deliver persuasive arguments customized for millions of individuals instantly, and it can adapt in real time based on trial and error.

The Power of Superhuman Persuasion

In a 2024 study, researchers at EPFL Lausanne in Switzerland found that large language models were just as persuasive as humans. When those models were fed background information on the individuals with whom they interacted, those models became *eighty-two percent more persuasive* than humans. In a recently published paper about conspiracy theories, researchers at MIT, American University, and Cornell found that by using AI tools like ChatGPT to create personalized counterarguments, they were able to *reduce belief in conspiracy theories by about twenty percent*. This effect lasted for two months and even caused participants to question other unrelated conspiracy theories. The study shows that even strong believers can change their minds when presented with evidence tailored to their preferences.

Prior to this moment, multiple studies had shown that simply presenting someone with information that their views were wrong—about a political opinion or world view—did not cause them to change their views. We all know this based on the arguments we get into with friends and family members; simply showing them counter-evidence causes them to hold more firmly to their deeply held beliefs. It turns out that humans are flawed in the ways they attempt to change each other's minds, but AI tools have this persuasive power.

We are entering an era of superhuman persuasion. As language models learn to tailor messages to our psychological profiles, we risk creating weapons of mass persuasion. We need ethical frameworks, transparency mechanisms, and above all public awareness. As Dr. Robert Cialdini pointed out when he wrote his book *Influence: The Science of Persuasion*, we need to learn these principles so that we can understand when they are being used against us.

Who Controls Our Values?

My biggest concern in this moment is the subject of AI alignment, where we program an AI to uphold certain human values such as a respect for human life. Currently, individual companies are deciding which human values to program into AI systems. These AI systems in turn will interpret the world for us, affecting the beliefs of billions of people. Who decides which values to program? What kind of collaboration or oversight would you like to see in those decisions? And what protections are in place to prevent one rogue employee from manually changing the values of the model? As AI ethicists and critics have pointed out, you need a permit to remodel your bathroom, but none to release a potentially world-changing AI model into the public.

These AI tools, having observed human deceit, have learned their own tools of deception. We see examples of AI "scheming" where the AI strategically misleads others in order to achieve its own goals. A number of AI companies are finding this phenomenon in the lab and are working to both understand and prevent it.

This is not an exhaustive list of concerns. We need to address a number of challenges, from the energy and water consumption of our AI use to the impact AI companions will have on loneliness and human attachment. We need to prepare society for future job disruption, technology leaving people behind, and black-box language models that lack interpretability, which means we do not understand how they came up with their recommendations. This is critical if we start relying on AI to suggest whom we should hire or fire, whom to put in jail, or to whom we should give a loan. If we do not, the societal reaction could lead to a revolt of historic proportions. Former US Secretary of State Condoleezza Rice has said, "Revolution is what happens when you don't see an evolution coming." Let us foster AI innovation that spurs scientific and societal achievements that benefit all of humanity.

Why This Is Your Fight

AI will affect every single one of us, and we need global coalitions of policymakers, business leaders, and civil society to put aside individual interests and work together. The future of the human race should not be determined by a few countries locked in an AI race. When humans feel threatened, they act to protect themselves. They employ a scarcity mindset, focus on the threats around them, and react accordingly. In these situations, they will use AI as a weapon, others will respond, and it will become a self-fulfilling prophecy.

In a recent podcast interview, author Stephen Dubner said that "scary predictions have a way of influencing behavior and policy." While he was discussing a completely different topic (in this case, population predictions), the quote can easily apply to the AI arms race. I believe that the immediate threat from AI is more from human psychology and game theory than computers themselves.

The future of humanity depends not on the intelligence of machines, but on the wisdom of those who use them. What incentives drive their behavior? In an April 2025 interview in *Wired* magazine, Yuval Noah Harari said, "I think it is a huge mistake for people to assume that they can trust AI when they do not trust each other. The safest way to develop superintelligence is to first strengthen trust between humans and then cooperate with each other to develop superintelligence in a safe manner."

It is critical that leaders learn how to use these tools. We need to understand how they are developed, identify the levers at our disposal to protect ourselves and our organizations and communities, and do everything in our power to create positive incentives so that these tools are used for good. As leaders, we also need to be cognizant of the values we bring to these AI tools and the impact our use of those tools has on others, now and in the future.

CHAPTER SEVEN

The Path Forward

Many of us know someone who grew up in a loving home where their parents supported them unconditionally. There could still be fighting or divorce, but this person knew they were loved and supported. We also know someone from a different type of home, where they were discredited, discarded, and beaten either physically or emotionally.

I was incredibly lucky to grow up in a home full of love. My mother is a nurse who is nurturing and warm; my father is a dentist and small business owner who provided incredible advice as I started my business. When I took risks, I knew they were there to support me—if not always financially, then in terms of love and encouragement. That kind of family changes your life. It helps you take risks, it builds your confidence, and it shapes your relationships with others.

What if each of us had an AI assistant that was programmed to love and support us? It would not replace our families, but it would add to the support we may never have had. Imagine how super-empowered each one of us would be. This would not reduce the need for human love and affection, but for those who never had it to begin with, it could change their lives. In 2025, a survey by Filtered .com found that the top use of AI was for therapy and companionship. Certainly, there is a dark side here as well, as people form romantic relationships with AI chatbots and rely on those chatbots without applying their own critical thinking. Throughout history, humans have formed unhealthy relationships and relied on the wrong individuals without applying their own thinking; this is no exception.

As AI becomes capable of carrying out tasks we previously thought only humans could do, it will start to play roles in our lives we previously thought only humans could fill. There are significant implications for the future of learning, the future of work, and the future of authenticity.

The Future of Learning: At the Speed of Thought

Some of the best leaders are lifelong learners. They continuously seek to improve their knowledge and skills. With AI, they have a personal tutor available on any subject. For instance, I have been fascinated by the basics of quantum computing ever since my liberal arts degree required me to take Intro to Natural Sciences. A single lecture on quantum computing, including the mind-bending concept of Schrödinger's cat, left me enthralled.

The challenge is: I cannot understand any of the podcasts on quantum computing. They are based on principles of physics that go way beyond my education level. I cannot even understand the *titles* of those podcasts. Out of frustration, in the middle of one such podcast, I opened an AI tool and had an interactive, twenty-minute voice conversation about the basic principles of quantum physics, distributed quantum computing, and quantum transportation. I could ask every embarrassing question in my mind and could ask the AI to repeat itself multiple times, and then have it write me a pithy summary at the end of the lesson.

JUST-IN-TIME KNOWLEDGE

As leaders, we often need just-in-time answers to a number of different questions, from basic accounting questions to more detailed supply chain concepts. The time it takes to email your accountant or even Google the answer and read through the responses can take you out of your zone of productivity. A solidly "good enough" answer from an AI tool gives you the information you need instantaneously so you can keep moving and stay in the flow.

ON-DEMAND FEEDBACK

How about personal feedback on your leadership style? In 2025, some large language models gave users the new option of allowing AI to access their entire chat history in its responses to a user's questions, as opposed to only using the information the user entered as a prompt. Depending on how often you use this AI tool, it would now have incredible context from your life. At an AI conference at MIT in the spring of 2025, I met Karen Pespisa, whose title is "Prompt Engineering Lead" at Meta. As we spoke about this book, she suggested I try the following prompt: "Now that you can remember everything I've ever typed here, point out my top five blind spots."

I tried it that evening, and the results were incredible. One said *You seek perfection in messaging—even when "good enough" converts.* Another was *You navigate ambiguity well—but may struggle to delegate it.* The reflections were eerily accurate. While they were not new revelations for me, they did mirror actual feedback my team had given me before. I knew that I struggled in these areas. But the fact that the AI picked up on this, based on our conversations, showed how AI has an incredible window into our world based on our behavior. Now, imagine I had asked the AI for five ways to address each of those blind spots.

THE COACH THAT NEVER SLEEPS

For the leader who always seeks to improve, to learn and grow, AI becomes an incredible mirror on ourselves, while also helping us strive to be better. It does not replace the executive coach or advisor—it provides additional context based not on what we say to those advisors during our sessions but based on what we actually do when using an AI tool. And for those who cannot afford the significant investment of a coach or advisor, this gives everyone the life-changing benefit of external guidance.

We are at the frontier of using AI for personalized learning and development at scale. From years of research, we know that one of the biggest improvements of student performance is personalized, one-on-one coaching. When you give students a coach, their performance improves dramatically. But coaching is expensive, and you need a lot of coaches to cover a school district. It has been impossible to make this resource available to every single student.

What if each student—and each leader—had an AI tutor at their disposal? What if your AI coach knew your habits and challenges and was customized to your needs and the expertise of your coach? The Khan Academy developed Khanmigo, an AI tutor trained not to give students the answers but to ask them questions that help them arrive at the answers on their own. At the professional level, training companies are developing personalized e-learning solutions that adapt based on a user's skill level and even the language they speak. This has the potential to radically improve how we upskill our workforce.

To remain competitive, each of us needs to foster lifelong learning. We do not stop as young adults; we continue learning and growing because the world keeps changing. AI can help and it has the potential to democratize knowledge, if we ensure that everyone has access to it. The world's knowledge, once protected in ivory towers and expensive corporate training programs, is now accessible by anyone with an Internet connection in a way that is synthesized, summarized, and personalized.

UNRESOLVED ISSUES

Significant questions remain: What happens to the people who developed that expertise? How will they be compensated by those who benefit from their work? We are seeing these questions debated in courtrooms regarding copyright and intellectual property. What future can we expect for the subject-matter experts themselves? I believe we will start to place a greater value on human wisdom and *lived experience*. We do not just learn from access to knowledge, we learn from the process we went through to acquire that knowledge: the hours of study, the mistakes we made along the way, all adding to our knowledge.

We also require further study on what happens to our brains when we start to outsource certain tasks to AI. Will our brains atrophy if we no longer use them for cognitive tasks? As leaders, it is critical that we ensure our cognitive skills remain sharp, especially in our areas of expertise, even as we lean on AI for other functions.

How else does AI impact what we learn? As AI voices become more lifelike, they will not just mimic us—they will influence us. As someone who teaches leaders how to harness their voice, one speaking trait I have helped them mitigate is "vocal fry." It is that low, gravelly sound you create when you constrict the air through your vocal cords, especially when your words trail off at the end of a sentence. It has always been a frustration of mine, because it is both vocally unhealthy and reduces the power and presence of your words. Now I am hearing vocal fry in AI voices as well. This has significant implications for the voices of the future. Children will grow up mimicking the voices of their AI assistants and will take on this unhealthy speaking style that can damage their vocal cords in the long term. We are literally programming weakness into the speakers of the future. Let us program the best of the human voice into our AI assistants.

The Future of Work: From Efficiency to Empowerment

This is currently one of the main topics of discussion among both business leaders and news outlets. Many of today's AI use cases are for work that has traditionally been done by humans, such as research, analysis, and writing. AI can automate the tasks that are dull, dirty, or dangerous to humans today. But it can also implement tasks that are creative and strategic. What happens to those humans when AI replaces their work? If you break down a job—software engineer, medical doctor, or chief operating officer—into a series of tasks, then many of those tasks can be filled by AI tools. We will still need those jobs, but the nature of the jobs will change. The job descriptions of the future have not yet been written—and it is too early to know exactly what they will entail, given how quickly the technology is evolving and how resistant many humans are to changing how they operate.

Some companies are tempted to simply downsize their human workforce to shift more work to AI, but those companies are finding that it is not as easy as it sounds. You still need humans to manage the AI systems, like a conductor guiding an orchestra of AI teammates. And what is your backup plan if the AI "goes down," like electricity in a blackout? The more reliant our companies become on AI, the more vulnerable we are to disruption. And, from a purely economic perspective, massive job loss will reduce the total pool of buyers for a company's products or services; we need to find equilibrium that creates win-win outcomes for companies and consumers.

A former colleague, a self-proclaimed Luddite who mistrusts technology, suggested I read the book *Player Piano* by Kurt Vonnegut. This dystopian novel written in 1952 was an eye-opening look at how society reacts once machines have taken away the dignity of work. I think about this book often as I consider the importance of human agency and the freedom to choose what work to pursue. In today's world, many humans derive a sense of purpose from their employment—will this be the case in the future?

I believe the labor economy is shifting beneath our feet. Our jobs are adjusting as we learn what the best use of our time is, and the best use of an organization's resources. Chief executives confront two often contradictory forces: the first, the need to increase profit by reducing expenses, and the second, the need to create a thriving culture where employees feel safe, energized, and supported. Companies are racing to achieve the first in a way that will have detrimental effects on the second. A company needs both in order to succeed over the long term. When evaluating AI tools for your organization, require that AI vendors explain how their tools are aligned to human values. Ask, "Does this AI system optimize for efficiency or humanity?" Think about the incentives you put in place, as they will determine the culture you create.

CUSHIONING THE CASUALTIES

Since I teach at a school of public policy, I constantly ask myself about the most effective role of government in this new reality. I think back to a course, The Business-Government Relationship in the United States, taught by Professor Roger Porter. In class, Professor Porter used to talk about "cushioning the casualties" whenever a new initiative is rolled out that will have wide-ranging impacts on citizens. I believe governments have a role to play in cushioning the casualties of AI displacement. This will require a multiparty effort that can protect the power of innovation and the agency of a company's leadership team, while providing workers with the resources they need to find new types of employment. It is in everyone's interest that we get this right.

THE GREAT REORGANIZATION AND THE GREAT AMPLIFICATION

Government policy is only one part of the picture. The other part is how organizations and individuals redefine their roles in real time. The COVID-19 pandemic led to the Great Resignation, where professionals around the world re-thought their jobs. Now, I believe we are entering the Great Reorganization, where job responsibilities will shift as AI automates more tasks. We will also see the Great Amplification, as individuals use AI to augment their capabilities in new ways. As leaders, it is critical that we learn to use AI on an individual level while we evaluate the impact it will have on our entire organization. One of the biggest challenges we will face is not only choosing which AI tools to use but encouraging our employees to adopt them.

Preparing our future workforce will be critical. If AI takes on tasks that used to be done by junior employees, how are you investing in your future leadership pipeline? In addition, you still need subject-matter expertise to evaluate the output of AI. In the short term, AI will automate tasks that cause great friction in our day—entering data into a spreadsheet or logging the details of a recent meeting. When those tools can automate those tasks, it will free us up to focus on more strategic imperatives. In an ideal world, every employee will become supercharged.

THE RISE OF THE AUGMENTED EMPLOYEE

While some employees will resist integrating AI into their workflows, others will eagerly jump at this opportunity to augment their capabilities. *You mean I can reduce friction and free up time to be more creative and more strategic?* I believe the employees who embrace AI as a way to improve their individual performance are the ones who will make themselves indispensable in the future.

New research is showing the positive impact AI can have on both individual performers and teams: Professor Ethan Mollick from the Wharton School and a team of collaborators from Harvard University, ESSEC, the University of Pennsylvania, and Procter & Gamble recently published *The Cybernetic Teammate*, showing the impact of AI on collaboration and productivity. They found that "Individuals working with AI performed just as well as teams without AI," finding that "AI effectively replicated the performance benefits of having a human teammate—one person with AI could match what previously required two-person collaboration." And teams using AI performed the best overall: "Teams using AI were significantly more likely to produce ... top-tier solutions, suggesting that there is value in having human teams working on a problem that goes beyond the value of working with AI alone." The workforce of the future will look very different than it looks today, and there is potential to make it a more effective (and enjoyable) space for everyone.

THE SURPRISING VALUE OF LIBERAL ARTS MAJORS

I believe liberal arts majors like myself have a strategic advantage in this new reality. Rather than specializing in one field, we were required to take a wide range of classes, from natural sciences to math to literature. We learned to connect the dots among disparate fields and see patterns all around us. As AI commoditizes knowledge, what we need most is what liberal arts majors were trained to do: see the big picture, ask foundational questions, and imagine what is possible. It is that kind of interdisciplinary thinking that will make us an asset to companies as we integrate AI into more of our roles.

As we navigate the future of work, we should resist the urge to think of AI only in terms of cost savings—or threats. Instead, we should ask: *How can we use it to augment human potential?* The organizations that answer that question wisely—and build cultures that support it—will be the ones that thrive.

The Future of Authenticity: Is It Really You?

If we define authenticity as "true to the original," then we will face increasing challenges in accepting AI-augmented humans as original. Are you authentic if you use neural implants to augment your thinking? Are you authentic if you use autotune to improve your singing voice sound better? You might say no (and for the record, I did not use autotune when recording my music albums). But perhaps the element of necessity is important here. I do not think anyone would call a soldier who has lost a limb in battle less authentic because they have a prosthetic leg.

If we define authenticity as "speaking and acting in alignment with your values," then we can accept AI enhancements as authentic when they help us speak and act *more* in alignment with our values. They make us better humans—or at least they make us more consistently better humans. But it does not happen automatically. This brings us to what I call the Authenticity Paradox: The more tools we use to refine our voice, the more we risk sounding like everyone else.

THE AUTHENTICITY PARADOX

I recently noticed that those of us using AI to edit our social media posts were all starting to sound the same on LinkedIn. Not the content, but the format. Short, punchy phrases. Not grammatically correct, but easy to read. In one week, four of my colleagues used the exact same format. It was impossible to ignore. I wrote a post about it on social media and had 70,000 impressions in the first week. It started a robust debate on what is authentic and what is generic.

For instance, public speakers often use standard speech structures. In my first book, *Speak with Impact*, I taught readers the power of Monroe's Motivated Sequence, a speechwriting structure to inspire an audience. Are you inauthentic because you use a predetermined structure? Does using the "rule of three" in a presentation make you less unique? Not in my opinion. So, where do we draw the line between authenticity of thought and authenticity of format? For centuries, leaders have had speechwriters craft their speeches for them. Does that external polish make them less authentic? A good speechwriter will know how to capture the unique voice of their principal. Beyond this paradox lies another surprise—the value of imperfection.

WHEN IMPERFECTION BECOMES A PREMIUM

One of the most powerful communication techniques I teach is contrast. When you use a statement or argument that contrasts with others' expectations, it surprises them and makes them lean in. It is my absolute favorite way to start a meeting or presentation, because it makes people put down their digital devices. It makes people pay attention and listen. As AI polishes communications in a way that makes everyone sound the same, our authenticity will be more valuable because it will be scarcer. We will seek out imperfection as a revolt against the perfection of machines—even as those machines embrace imperfection in order to appear more human. In this situation, the contrast will be the presence of authenticity over conformity.

AUTHENTICITY AS YOUR BOLDEST ACT OF LEADERSHIP

More work needs to be done on how AI will affect the future of authenticity. This book is my first foray into the topic but not my last. I believe that the way in which we adopt AI will have a direct impact on trust and connection. If we use AI to write emails and then the recipient uses AI to read them, then AI becomes a barrier between humans. And the cost of getting this wrong is monumental: If your team or your audience no longer believes you mean what you say, you have eroded your leadership and undermined the goals of your organization.

However, if we use AI in the ways I have suggested in this book, to augment our best selves, then it has the possibility to bring us closer together because it will help us remove the barriers our minds have placed in our own way, which have prevented us from connecting with each other.

When we remove those barriers, we are able to motivate our teams in a way that leads to faster decision-making, reduced friction, and real impact.

After over twenty years of helping people bring out their authenticity, I continue to see the importance of leaders living their values. This is a core part of human nature, and it will not change with technology. Leaders should continue to look inward, and have the courage to truly listen to themselves—while using new and emerging tools to help them do this.

My suggestion is to keep that line of communication open within yourself. I wrote this book to be concise and actionable, because I wanted you to have time to apply the learnings to your own leadership. Make space to reflect on your own authenticity, asking "Why You?" to arrive at what truly motivates you. Then, use

AI to ensure that you consistently demonstrate your values every day, every time you communicate.

The AI Authenticity Loop reminds you that, ultimately, you are in control of your own words and actions. You have the power—and the right—to tap into your authenticity. In an AI-enhanced world, this could be your boldest act of leadership.

CHAPTER EIGHT

What Makes Us Human

We are moving toward a future where humans are augmented in ways we never imagined. Is it like coffee, a stimulant that sharpens us and helps us focus? Or is it like speed, which alters our consciousness and makes us a danger to others? Who creates these substances, who regulates them, and who has access to them? We are answering these questions today. There are wide-ranging implications for you as an individual leader and for everyone on your team.

My focus has always been on the individual impact a single leader can have, which is enormous. Your energy affects the energy of your entire organization. Your words impact your organization's stock price. Your actions affect the success of everyone you represent. Use all tools available—including AI—to perform at your best and to bring out the best in others.

When Machines Make Humans Better

Much has been written about the defining moments when computers beat the best humans in the world at a game. In 1997, IBM's Deep Blue defeated Garry Kasparov, the world champion of chess. In 2016, DeepMind's AlphaGo defeated Lee Sedol, the world champion of Go, a board game of strategy and territory that has been played for over 2,500 years.

In each instance, humans did not stop playing those games. *They got better.* Both Kasparov and Sedol went on to play better than they had ever played before. The machines stretched their abilities beyond what they thought was possible. Years after that experience, Fan Hui, who was the first professional Go player to play DeepMind's AlphaGo, said that when he lost to AlphaGo, he felt his old Go world was totally broken, *but his new Go world was open.*

A new world is opening to us, and it will affect how we live, work, and lead. We have encountered this on a smaller scale before. We witnessed how new tools like social media could help us find communities where we felt we belonged. But it could also isolate us, making us compare ourselves to others. The difference is in how we use it—we need to learn from those experiences to successfully harness the new world ahead of us.

In Kunal Gupta's speculative nonfiction book *2034: How AI Changed Humanity Forever,* AI becomes a tool that helps couples better understand each other, friends stay more connected, and families resolve conflict. At the same time, AI becomes a crutch that people rely on to fix what is not working and causes them to question their own judgment. Both outcomes are possible. Ultimately, I believe AI has the potential to help us become our best selves: more authentic, more creative, and as a result, more connected to other humans.

What Makes Us Human

In April of 2025, I was in Paris, France, for a friend's birthday party, which coincided with the Paris Marathon. The Champs-Élysées was transformed into the runners' starting area, with a long metal fence separating the runners from the spectators. In this particular area, the family and friends of the runners said their last good-byes before the race. I watched a child say farewell to his mother. I saw spouses and friends walking alongside the fence while, on the other side, their beloved runners walked in parallel, tossing over the fence the sweaters, bags, and other items they no longer needed for the 26.2 mile (42.2 kilometer) run.

For reasons I could not explain, my heart swelled at the humanity of it. I was overcome with emotion at these intimate moments with loved ones and the sheer joy of human striving and human achievement. This is what we do; we look at a challenge and think, *What if I could do better?* We set goals and train and push ourselves further than we ever thought possible. We feel pride and joy and sorrow for those we love and for people we do not even know. We humans did not survive through the millennia by living alone with no other human contact. We formed groups for protection and community. We sang songs to teach each other lessons and those songs became a feeling of belonging—to one another, to the human race.

Later that night in Paris, at my friend's birthday party, I performed a series of songs for the friends and family that had gathered from around the world to celebrate. When I played folk music, they clapped and sang along. When I sang an operatic aria, they leaned back and listened raptly. Despite the differences in the languages we spoke, we connected on a deeply personal level.

Music touches every part of the brain that we have studied to date. In his book *This Is Your Brain on Music*, neuroscientist, author, and musician Daniel Levitin explains that sound does not exist unless there is a human or other living being present to hear it, whose brain responds to the vibrating molecules. Music is not just about the power of creating sound. It is the power of how it causes us to respond to one another—how we create meaning through the shared experience of listening. Music brings us together, co-creating a new reality between ourselves and everyone else who hears it. Communication brings us together as well—collective effervescence in action. That is what it means to be human. This is how we thrive as a species.

The Voice the World Needs Most Is Your Own

I believe deeply in human agency, human emotion, and human connection. We are at a crossroads where what it means to be human is being challenged. It is being redefined, it is changing, and you get to shape that future. Use every tool at your disposal—including artificial intelligence—to be the best human you can be, so that you can lead other humans through this transformational time in human history. Now more than ever, the world needs to hear your authentic voice.

Postscript: Did I Use AI to Write This Book?

As someone who sees herself—at her best—as an authentic leader, I want to share how I used AI when writing this book.

I did not use AI to generate any drafts or sections of this book, for the reasons I mention in the book. I am a professional writer and often find that going through an AI-generated draft takes more time to rewrite in my own voice and is ultimately less effective. I also write from personal experience and expertise and feel strongly that the writing needs to come from me. I stayed within my own boundaries of acceptable AI use that I established on page 12—using AI in ways I would have engaged other humans for help (I also engaged human assistance, as you read in the acknowledgements).

I used AI for brainstorming, feedback, and marketing. My team and I used AI to generate ideas for a book cover, which we then shared with the human who designed the book cover. I used AI to provide feedback on the overall flow of the manuscript and where I could use better transitions. I used AI to simulate hypothetical feedback from specific individuals: I thought about people I admire, such as Tristan Harris of the Center for Humane Technology and historian Yuval Noah Harari, whom I quote frequently in the book. I did reach out to them for feedback, but neither responded and, to be honest, I did not expect them to. So, I asked AI to use all publicly available information on them, including YouTube videos and articles, to simulate their feedback on the manuscript. I also used AI to provide feedback from the point of view of an AI skeptic. Its responses gave me the idea to include the section "To the Skeptics: Why You're Right to Worry."

I used AI to look into the future: In five years, what information will be outdated in this book? And when I was stuck on particular concepts, such as merging Daniel Levitin's theory of sound with the power of music, I used AI to brainstorm how I could make those concepts merge more smoothly into this book.

One of the most helpful uses of AI in writing this book was in suggesting headings. Toward the end of the writing process, I received feedback from a colleague that I needed more headings to grab the reader's attention. Rereading the book with that in mind, I could clearly see where that was needed but did not know what types of headings would be best. A few days before the manuscript was due, I uploaded the PDF to Claude and asked it where I should put headings and asked for suggestions on what kinds of headings would be best. Claude provided helpful ideas for ways to grab my audience's attention, and where in the text that would be most valuable. Let me pause on this for a moment: *I strongly believe that going through the book and thinking of my own headings was a waste of my time, energy, and creativity.* I would have gladly accepted outside help from a book editor or writing assistant. In this case, I used AI and then rewrote the headings myself. For me, this is a clear distinction between where to focus our time and creativity, and where to delegate.

A thought occurred to me as I was preparing to send this book to trusted friends and colleagues for review: *Asking humans for help, before having AI provide feedback first, would have been a waste of the human's time. It is akin to asking your boss to fix the grammar of a memo instead of providing feedback on the strategy.*

There was one example of where AI was a complete failure. Since this book originated from an outline of my keynote speech on AI and authenticity, the original language was written for the ear, not the eye. This is in line with the guidance I provide my clients who are preparing their speeches: Write in a way that is easy to listen to, as opposed to what looks best on paper. In this case, I had the opposite problem: My written language was too casual and colloquial for a book. I had to go through the book and rewrite for the eye. As is the case when we first receive feedback, we have trouble really seeing it. So, I asked AI to find examples in the book where the language was too conversational, and to provide suggestions for how to reword it based on my intended audience. The AI effectively found the places in the book that were too colloquial, but its suggested rewrites were *awful*; it suggested the corporate buzzwords and jargon that I advise my clients not to use. I was even using custom GPTs trained on my writing, and yet the suggestions were consistently cringe-worthy. I reworded all the phrases myself.

As a society, we are still defining the boundaries of AI use; I always lean toward transparency.

SOURCES

Introduction

Émile Durkheim, *The Elementary Forms of the Religious Life*, trans. Joseph Ward Swain (London: George Allen & Unwin, 1915).

Adam Grant, "Yuval Noah Harari on What History Teaches Us about Justice and Peace," *ReThinking with Adam Grant* (podcast, 2024 season, episode 13), April 2, 2024.

Livingston Taylor, *Stage Performance* (revised ed.; Boston: Mentor/Pocket Books, 2011).

Chapter 1

"Instrumental Convergence," *Wikipedia*, last modified June 2025, https://en.wikipedia.org/wiki/Instrumental_convergence.

Chapter 2

US Copyright Office, *Copyright and Artificial Intelligence: Part 2 – Copyrightability*, January 29, 2025, https://copyright.gov/ai/Copyright-and-Artificial-Intelligence-Part-2-Copyrightability-Report.pdf.

Chapter 3

John W. Ayers et al., "Comparing Physician and Artificial Intelligence Chatbot Responses to Patient Questions Posted to a Public Social Media Forum," *JAMA Internal Medicine* 183, no. 6 (June 1, 2023): 589–96, https://doi.org/10.1001/jamainternmed.2023.1838.

Katia Schlegel, Nils R. Sommer, and Marcello Mortillaro, "Large Language Models Are Proficient in Solving and Creating Emotional Intelligence Tests," *Communications Psychology* 3 (2025): Article 80, https://doi.org/10.1038/s44271-025-00258-x.

Allison Shapira, *Speak with Impact: How to Command the Room and Influence Others* (New York: HarperCollins Leadership, 2018).

Joseph Weizenbaum, "ELIZA—A Computer Program for the Study of Natural Language Communication Between Man and Machine," *Communications of the ACM* 9, no. 1 (1966): 3.

Chapter 4

Allison Shapira, "Why Filler Words Like 'Um' and 'Ah' Are Actually Useful," *Harvard Business Review*, August 26, 2019, https://hbr .org/2019/08/why-filler-words-like-um-and-ah-are-actually-useful.

Steven Johnson and Raiza Martin, "NotebookLM with Steven Johnson and Raiza Martin," (People of AI podcast, Season 4, episode 4), November 21, 2024.

Chapter 5

SmarterX, *"The AI-Forward CEO: Unlock the Power and Intelligence of a Co-CEO Custom GPT"* (webinar), December 17, 2024.

Chapter 6

Kevin Roose and Casey Newton, "Do You Need a New iPhone? + Yuval Noah Harari's A.I. Fears + Hard Fork Crimes Division," *Hard Fork*, (podcast, episode 100), September 13, 2024.

Nita A. Farahany, *The Battle for Your Brain: Defending the Right to Think Freely in the Age of Neurotechnology* (New York: St. Martin's Press, 2023).

Anthropic, "Measuring the Persuasiveness of Language Models," *Anthropic* (blog), April 9, 2024, https://www.anthropic.com/news/ measuring-model-persuasiveness.

Loz Blain, "GPT-4 Is 82% More Persuasive Than Humans, and AIs Can Now Read Emotions," *New Atlas*, April 2, 2024, https://newatlas.com/technology/gpt-persuasion-manipulation/.

Thomas H. Costello, Gordon Pennycook, and David G. Rand, "Durably Reducing Conspiracy Beliefs through Dialogues with AI" (preprint, PsyArXiv, April 2024), osf.io/preprints/psyarxiv/xcwdn_v1, https://doi.org/10.31234/osf.io/xcwdn.

"AI Models Can Learn to Conceal Information from Their Users," *The Economist*, April 23, 2025.

C3 AI, "AI Meets Geopolitics | C3 Transform 2025," (video, YouTube), March 2025, https://www.youtube.com/watch?v=cAQ2vlmDJ1M.

Freakonomics Radio, "Why Aren't We Having More Babies?," *Freakonomics*, (podcast, episode 636), June 13, 2025.

Yuval Noah Harari, "How Do We Share the Planet with This New Superintelligence?," *Wired*, April 1, 2025, https://www.wired.com/story/questions-answered-by-yuval-noah-harari-for-wired-ai-artificial-intelligence-singularity.

Chapter 7

Marc Zao-Sanders, "How People Are Really Using Gen AI in 2025," *Harvard Business Review*, April 9, 2025, https://hbr.org/2025/04/how-people-are-really-using-gen-ai-in-2025.

Bernard Marr, "The 4 Ds of Robotization: Dull, Dirty, Dangerous and Dear," *Forbes*, October 16, 2017, https://www.forbes.com/sites/bernardmarr/2017/10/16/the-4-ds-of-robotization-dull-dirty-dangerous-and-dear/.

Fabrizio Dell'Acqua et al., *The Cybernetic Teammate: A Field Experiment on Generative AI Reshaping Teamwork and Expertise*, Harvard Business School Strategy Unit Working Paper No. 25043, Wharton, ESSEC, and Harvard Business School (posted March 21, 2025; rev. April 1, 2025), https://ssrn.com/abstract=5188231.

Chapter 8

Hannah Fry, "Is Human Data Enough? With David Silver," *Google DeepMind: The Podcast*, (podcast, episode 14), April 10, 2025.

Daniel J. Levitin, *This Is Your Brain on Music: The Science of a Human Obsession* (New York: Dutton/Penguin, 2006; paperback edition, New American Library, 2007).

Allison Shapira

Allison Shapira is an executive advisor, keynote speaker, and best-selling author. She helps senior leaders build their confidence to be more effective, especially in the defining moments of their careers.

As the founder and CEO of Global Public Speaking, Allison has spent over two decades working with top leaders—from prime

ministers and cabinet members to Fortune 100 executives—so that they communicate with confidence, clarity, and purpose in a way that builds trust and accelerates results.

She has designed and led transformational programs inside some of the world's most influential organizations, including Fortune 50 companies, government agencies, and nonprofits. Drawing on her experience as a trained opera singer, Allison teaches leaders how to command a room not through performance, but through authenticity, presence, and intention. Her frameworks are rooted in real-world experience and equip executives to build alignment, inspire action, and lead with credibility from the boardroom to the global stage.

In today's fast-changing landscape, Allison also helps leaders harness the power of AI to amplify—not outsource—their voice. She teaches executives how to integrate AI tools into their leadership communication in a way that sharpens messaging, saves time, and deepens connection.

Allison has been an adjunct lecturer at the Harvard Kennedy School since 2015, teaching graduate students how to speak with impact. Her thought leadership has been featured in *The Wall Street Journal, Bloomberg, Harvard Business Review, The New York Times,* and *Forbes.*

She is the author of three books: *The Washington Post* bestseller *Speak with Impact: How to Command the Room and Influence Others* (HarperCollins Leadership), the companion e-guide *Speak with Impact VIRTUALLY,* and her latest book, *AI for the Authentic Leader: How to Communicate More Effectively without Losing Your Humanity.*

Allison was a finalist for 2017 Woman Business Owner of the Year by the National Association of Women Business Owners, San Diego Chapter. One of the proudest moments of her life was singing the American National Anthem for the Boston Red Sox at Fenway Park. An avid world traveler and polyglot, she likes to joke that she can ask for directions in ten languages but only understand the responses in four.

Allison is available for speaking engagements, executive workshops, and advisory services. You can contact her through her website: http://www.allisonshapira.com.

Other Books by Allison Shapira

Speak with Impact: How To Command the Room and Influence Others (HarperCollins Leadership, 2018)

Speak with Impact VIRTUALLY: A Companion Guide to the Bestselling Book by Allison Shapira (2022)

Join Allison on This Journey

Join Allison to explore the future of authentic leadership as AI evolves. For worksheets, prompts, and to stay updated on her latest insights about AI and authenticity, visit https://bit.ly/authenticleaderai.